Undergraduate Texts in Mathematics

Fred H. Croom

Basic Concepts of Algebraic Topology

Springer-Verlag
New York Heidelberg Berlin

Fred H. Croom
The University of the South
Sewanee, Tennessee 37375
USA

AMS Subject Classifications: 55-01

Library of Congress Cataloging in Publication Data

Croom, Fred H 1941–
Basic concepts of algebraic topology.
(Undergraduate texts in mathematics)
Bibliography: p.
Includes index.
1. Algebraic topology. I. Title.
QA612.C75 514.2 77-16092

Printed in the United States of America.

9 8 7 6 5 4 3 2 1

ISBN 0-387-90288-0 Springer-Verlag New York
ISBN 3-540-90288-0 Springer-Verlag Berlin Heidelberg

Preface

This text is intended as a one semester introduction to algebraic topology at the undergraduate and beginning graduate levels. Basically, it covers simplicial homology theory, the fundamental group, covering spaces, the higher homotopy groups and introductory singular homology theory.

The text follows a broad historical outline and uses the proofs of the discoverers of the important theorems when this is consistent with the elementary level of the course. This method of presentation is intended to reduce the abstract nature of algebraic topology to a level that is palatable for the beginning student and to provide motivation and cohesion that are often lacking in abstact treatments. The text emphasizes the geometric approach to algebraic topology and attempts to show the importance of topological concepts by applying them to problems of geometry and analysis.

The prerequisites for this course are calculus at the sophomore level, a one semester introduction to the theory of groups, a one semester introduction to point-set topology and some familiarity with vector spaces. Outlines of the prerequisite material can be found in the appendices at the end of the text. It is suggested that the reader not spend time initially working on the appendices, but rather that he read from the beginning of the text, referring to the appendices as his memory needs refreshing. The text is designed for use by college juniors of normal intelligence and does not require "mathematical maturity" beyond the junior level.

The core of the course is the first four chapters—geometric complexes, simplicial homology groups, simplicial mappings, and the fundamental group. After completing Chapter 4, the reader may take the chapters in any order that suits him. Those particularly interested in the homology sequence and singular homology may choose, for example, to skip Chapter 5 (covering spaces) and Chapter 6 (the higher homotopy groups) temporarily and proceed directly to Chapter 7. There is not so much material here, however, that the instructor will have to pick and choose in order to cover something in every chapter. A normal class should complete the first six chapters and get well into Chapter 7.

No one semester course can cover all areas of algebraic topology, and many important areas have been omitted from this text or passed over with only brief mention. There is a fairly extensive list of references that will point the student to more advanced aspects of the subject. There are, in addition, references of historical importance for those interested in tracing concepts to their origins. Conventional square brackets are used in referring to the numbered items in the bibliography.

For internal reference, theorems and examples are numbered consecutively within each chapter. For example, "Theorem IV.7" refers to Theorem 7 of Chapter 4. In addition, important theorems are indicated by their names in the mathematical literature, usually a descriptive name (e.g., Theorem 5.4, The Covering Homotopy Property) or the name of the discoverer (e.g., Theorem 7.8, The Lefschetz Fixed Point Theorem.)

A few advanced theorems, the Freudenthal Suspension Theorem, the Hopf Classification Theorem, and the Hurewicz Isomorphism Theorem, for example, are stated in the text without proof. Although the proofs of these results are too advanced for this course, the statements themselves and some of their applications are not. Students at the beginning level of algebraic topology can appreciate the beauty and power of these theorems, and seeing them without proof may stimulate the reader to pursue them at a more advanced level in the literature. References to reasonably accessible proofs are given in each case.

The notation used in this text is fairly standard, and a real attempt has been made to keep it as simple as possible. A list of commonly used symbols with definitions and page references follows the table of contents. The end of each proof is indicated by a hollow square, $\square$.

There are many exercises of varying degrees of difficulty. Only the most extraordinary student could solve them all on first reading. Most of the problems give standard practice in using the text material or complete arguments outlined in the text. A few provide real extensions of the ideas covered in the text and represent worthy projects for undergraduate research and independent study beyond the scope of a normal course.

I make no claim of originality for the concepts, theorems, or proofs presented in this text. I am indebted to Wayne Patty for introducing me to algebraic topology and to the many authors and research mathematicians whose work I have read and used.

I am deeply grateful to Stephen Puckette and Paul Halmos for their help and encouragement during the preparation of this text. I am also indebted to Mrs. Barbara Hart for her patience and careful work in typing the manuscript.

FRED H. CROOM

Contents

List of Symbols

$\in$	element of 155
$\notin$	not an element of 155
$\subset$	contained in or subset of 155
$=$	equals
$\neq$	not equal to
$\varnothing$	empty set 155
$\{x: \ldots\}$	set of all x such that ... 155
$\cup$	union of sets 155
$\cap$	intersection of sets 155
$\bar{A}$	closure of a set 158
$X \setminus A$	complement of a set 155
$A \times B$, ΠX_α	product of sets 155, 157
$\lvert x \rvert$	absolute value of a real or complex number
$\lVert x \rVert$	Euclidean norm 161
$\mathbb{R}$	the real line 162
$\mathbb{R}^n$	n-dimensional Euclidean space 161
$\mathbb{C}$	the complex plane 69
B^n	n-dimensional ball 162
S^n	n-dimensional sphere 162
$(x_1, x_2, \ldots, x_n)$	n-tuple 155
$f: X \to Y$	function from X to Y 156
$gf: X \to Z$	composition of functions 156
$f\vert_C$	restriction of a function 157
$f(A)$	image of a set 156
$f^{-1}(B)$	inverse image of a set 160
$f^{-1}(y)$	inverse image of a point
f^{-1}	inverse function 156
$<$, $\leq$	less than, less than or equal to

List of Symbols

$>, \geq$	greater than, greater than or equal to
$\{0\}$	trivial group consisting only of an identity element 163
(a,b)	open interval
$[a,b]$	closed interval
I	closed unit interval $[0,1]$ 162
I^n	n-dimensional unit cube 106, 162
∂I^n	point set boundary of I^n 106, 162
X/A	quotient space 161
$\cong$	isomorphism 164
σ^n, τ^n	n-simplexes 8
$\dot{\sigma}$	barycenter of a simplex 46
$\|K\|$	polyhedron associated with a complex K; the geometric carrier of K 10
$K^{(1)}$	first barycentric subdivision of a complex K 47
$K^{(n)}$	nth barycentric subdivision of a complex K 47
$\langle v_0 \ldots v_n \rangle$	n-simplex with vertices $v_0, \ldots, v_n$ 9
$\mathrm{st}(v)$	star of a vertex 43
$\mathrm{ost}(v)$	open star of a vertex 43
$\mathrm{Cl}(\sigma)$	closure of a simplex 10
$[\sigma^p, \sigma^{p-1}]$	incidence number 13
$\alpha \sim x_0 \beta$	loops equivalent modulo x_0 61, 106
$B_p(\ )$	p-dimensional boundary group 18
$C_p(\ )$	p-dimensional chain group 16
$H_p(\ )$	p-dimensional homology group 19
$R_p(\ )$	pth Betti number 26
$Z_p(\ )$	p-dimensional cycle group 18
$\lambda(f)$	Lefschetz number of a map 136, 138
$\pi_1(\ )$	fundamental group 63
$\pi_n(\ )$	nth homotopy group 107
χ	Euler characteristic 27
$\partial(c_p)$	boundary of a p-chain 17
$\partial : C_p(K) \to C_{p-1}(K)$	boundary homomorphism on chain groups 17
$\partial : H_p(K/L) \to H_{p-1}(L)$	boundary homomorphism on homology groups 142
$g \cdot \sigma^p$	an elementary p-chain 16
Σ	sum
diam	diameter 159
dim	dimension 26
$A = (a_{ij})$	matrix 167
exp, e^z	the exponential function on the complex plane 69, 85
sin	the sine function
cos	the cosine function
$\oplus$	direct sum of groups 165
$\mathbb{Z}$	the additive group of integers 164

Geometric Complexes and Polyhedra 1

1.1 Introduction

Topology is an abstraction of geometry; it deals with sets having a structure which permits the definition of continuity for functions and a concept of "closeness" of points and sets. This structure, called the "topology" on the set, was originally determined from the properties of open sets in Euclidean spaces, particularly the Euclidean plane.

It is assumed in this text that the reader has some familiarity with basic topology, including such concepts as open and closed sets, compactness, connectedness, metrizability, continuity, and homeomorphism. All of these are normally studied in what is called "point-set topology"; an outline of the prerequisite information is contained in Appendix 2.

Point-set topology was strongly influenced by the general theory of sets developed by Georg Cantor around 1880, and it received its primary impetus from the introduction of general metric spaces by Maurice Frechet in 1906 and the appearance of the book *Grundzüge der Mengenlehre* by Felix Hausdorff in 1912.

Although the historical origins of algebraic topology were somewhat different, algebraic topology and point-set topology share a common goal: to determine the nature of topological spaces by means of properties which are invariant under homeomorphisms. Algebraic topology describes the structure of a topological space by associating with it an algebraic system, usually a group or a sequence of groups. For a space X, the associated group $G(X)$ reflects the geometric structure of X, particularly the arrangement of the "holes" in the space. There is a natural interplay between continuous maps $f: X \to Y$ from one space to another and algebraic homomorphisms $f^*: G(X) \to G(Y)$ on their associated groups.

Consider, for example, the unit circle S^1 in the Euclidean plane. The circle has one hole, and this is reflected in the fact that its associated group is generated by one element. The space composed of two tangent circles (a figure eight) has two holes, and its associated group requires two generating elements.

The group associated with any space is a topological invariant of that space; in other words, homeomorphic spaces have isomorphic groups. The groups thus give a method of comparing spaces. In our example, the circle and figure eight are not homeomorphic since their associated groups are not isomorphic.

Ideally, one would like to say that any topological spaces sharing a specified list of topological properties must be homeomorphic. Theorems of this type are called *classification theorems* because they divide topological spaces into classes of topologically equivalent members. This is the sort of theorem to which topology aspires, thus far with limited success. The reader should be warned that an isomorphism between groups does not, in general, guarantee that the associated spaces are homeomorphic.

There are several methods by which groups can be associated with topological spaces, and we shall examine two of them, *homology* and *homotopy*, in this course. The purpose is the same in each case: to let the algebraic structure of the group reflect the topological and geometric structures of the underlying space. Once the groups have been defined and their basic properties established, many beautiful geometric theorems can be proved by algebraic arguments. The power of algebraic topology is derived from its use of algebraic machinery to solve problems in topology and geometry.

The systematic study of algebraic topology was initiated by the French mathematician Henri Poincaré (1854–1912) in a series of papers[1] during the years 1895–1901. Algebraic topology, or analysis situs, did not develop as a branch of point-set topology. Poincaré's original paper predated Frechet's introduction of general metric spaces by eleven years and Hausdorff's classic treatise on point-set topology, *Grundzüge der Mengenlehre*, by seventeen years. Moreover, the motivations behind the two subjects were different. Point-set topology developed as a general, abstract theory to deal with continuous functions in a wide variety of settings. Algebraic topology was motivated by specific geometric problems involving paths, surfaces, and geometry in Euclidean spaces. Unlike point-set topology, algebraic topology was not an outgrowth of Cantor's general theory of sets. Indeed, in an address to the International Mathematical Congress of 1908, Poincaré referred to point-set theory as a "disease" from which future generations would recover.

Poincaré shared with David Hilbert (1862–1943) the distinction of being the leading mathematician of his time. As we shall see, Poincaré's geometric

[1] The papers were *Analysis Situs*, *Complément à l'Analysis Situs*, *Deuxième Complément*, and *Cinquième Complément*. The other papers in this sequence, the third and fourth complements, deal with algebraic geometry.

insight was nothing short of phenomenal. He made significant contributions in differential equations (his original specialty), complex variables, algebra, algebraic geometry, celestial mechanics, mathematical physics, astronomy, and topology. He wrote thirty books and over five hundred papers on new mathematics. The volume of Poincaré's mathematical works is surpassed only by that of Leonard Euler's. In addition, Poincaré was a leading writer on popular science and philosophy of mathematics.

In the remaining sections of this chapter we shall examine some of the types of problems that led to the introduction of algebraic topology and define polyhedra, the class of spaces to which homology groups will be applied in Chapter 2.

1.2 Examples

The following are offered as examples of the types of problems that led to the development of algebraic topology by Poincaré. They are hard problems, but the reader who has not studied them before has no cause for alarm. We will use them only to illustrate the mathematical climate of the 1890's and to motivate Poincaré's fundamental ideas.

1.2.1 *The Jordan Curve Theorem and Related Problems*

The French mathematician Camille Jordan (1858–1922) was first to point out that the following "intuitively obvious" fact required proof, and the resulting theorem has been named for him.

Jordan Curve Theorem. *A simple closed curve C (i.e., a homeomorphic image of a circle) in the Euclidean plane separates the plane into two open connected sets with C as their common boundary. Exactly one of these open connected sets (the "inner region") is bounded.*

Jordan proposed this problem in 1892, but it was not solved by him. That distinction belongs to Oswald Veblen (1880–1960), one of the guiding forces in the development of algebraic topology, who published the first correct solution in 1905 [55].

Lest the reader be misguided by his intuition, we present the following related conjecture which was also of interest at the turn of the century.

Conjecture. *Suppose D is a subset of the Euclidean plane $\mathbb{R}^2$ and is the boundary of each component of its complement $\mathbb{R}^2 \backslash D$. If $\mathbb{R}^2 \backslash D$ has a bounded component, then D is a simple closed curve.*

This conjecture was proved false by L. E. J. Brouwer (1881–1966) at about the same time that Veblen gave the first correct proof of the Jordan Curve Theorem. The following counterexample is due to the Japanese geometer Yoneyama (1917) and is known as the Lakes of Wada.

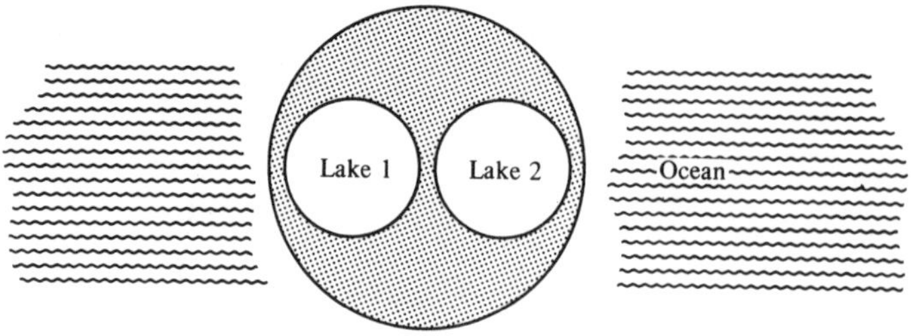

Figure 1.1

Consider the double annulus in Figure 1.1 as an island with two lakes having water of distinct colors surrounded by the ocean. By constructing canals from the ocean and the lakes into the island, we shall define three connected open sets. First, canals are constructed bringing water from the sea and from each lake to within distance $d = 1$ of each dry point of the island. This process is repeated for $d = \frac{1}{2}, \frac{1}{4}, \ldots, (\frac{1}{2})^n, \ldots$, with no intersection of canals. The two lakes with their canal systems and the ocean with its canal form three regions in the plane with the remaining "dry land" D as common boundary. Since D separates the plane into three connected open sets instead of two, the Jordan Curve Theorem shows that D is not a simple closed curve.

1.2.2 *Integration on Surfaces and Multiply-connected Domains*

Consider the annulus in Figure 1.2 enclosed between the two circles H and K.

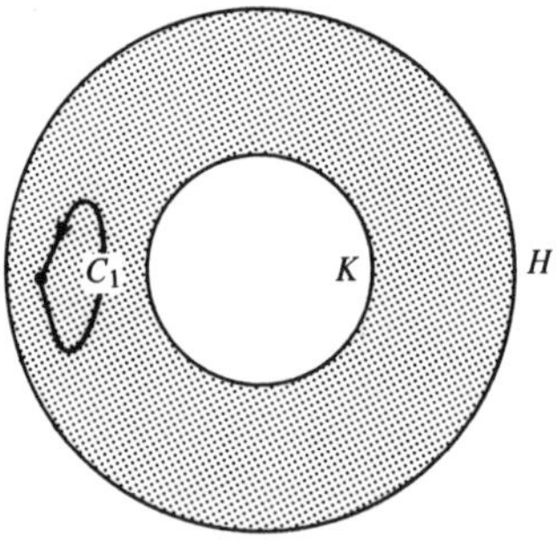

Figure 1.2

We are interested in evaluating curve integrals

$$\int_C p\,dx + q\,dy$$

where $p = p(x, y)$ and $q = q(x, y)$ are continuous functions of two variables whose partial derivatives are continuous and satisfy the relation

$$\frac{\partial p}{\partial y} = \frac{\partial q}{\partial x}.$$

Since curve C_1 can be continuously deformed to a point in the annulus, then

$$\int_{C_1} p\,dx + q\,dy = 0.$$

Thus C_1 is considered to be negligible as far as curve integrals are concerned, and we say that C_1 is "equivalent" to a constant path.

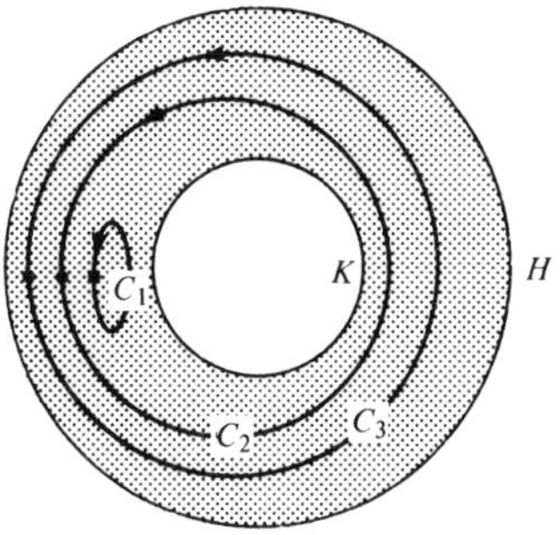

Figure 1.3

Green's Theorem insures that the integrals over curves C_2 and C_3 of Figure 1.3 are equal, so we can consider C_2 and C_3 to be "equivalent."

How can we give a more precise meaning to this idea of equivalence of paths? There are several possible ways, and two of them form the basic ideas of algebraic topology. First, we might consider C_2 and C_3 equivalent because each can be transformed continuously into the other within the annulus. This is the basic idea of homotopy theory, and we would say that C_2 and C_3 are *homotopic paths*. Curve C_1 is homotopic to a trivial (or constant) path since it can be shrunk to a point. Note that C_2 and C_1 are not homotopic paths since C_2 cannot be pulled across the "hole" that it encloses. For the same reason, C_1 is not homotopic to C_3.

Another approach is to say that C_2 and C_3 are equivalent because they form the boundary of a region enclosed in the annulus. This second idea is the basis of homology theory, and C_2 and C_3 would be called *homologous paths*. Curve C_1 is *homologous to zero* since it is the entire boundary of a region enclosed in the annulus. Note that C_1 is not homologous to either C_2 or C_3.

The ideas of homology and homotopy were introduced by Poincaré in his original paper *Analysis Situs* [49] in 1895. We shall consider both topics in some detail as the course progresses.

1.2.3 *Classification of Surfaces and Polyhedra*

Consider the problem of explaining the difference between a sphere S^2 and a torus T as shown in Figure 1.4. The difference, of course, is apparent: the sphere has one hole, and the torus has two. Moreover, the hole in the sphere is somehow different from those in the torus. The problem is to explain this difference in a mathematically rigorous way which can be applied to more complicated and less intuitive examples.

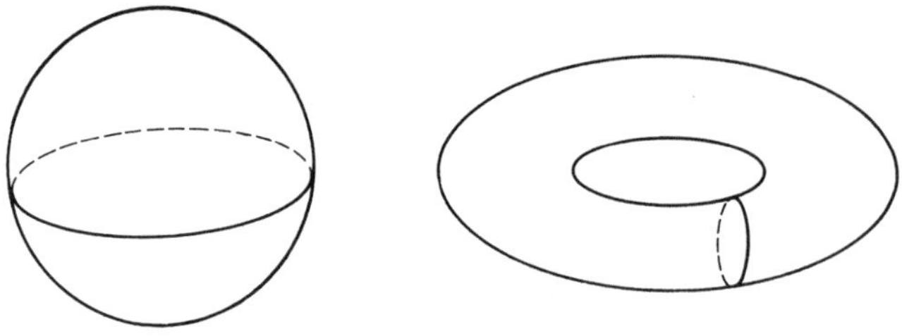

Figure 1.4

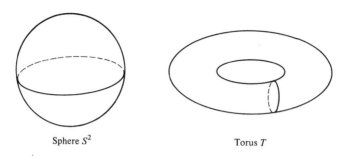

Consider the idea of homotopy. Any simple closed curve on the sphere can be continuously deformed to a point on the spherical surface. Meridian and parallel circles on the torus do not have this property. (These facts, like the Jordan Curve Theorem, are "intuitively obvious" but difficult to prove.)

From the homology viewpoint, every simple closed curve on the sphere is the boundary of the portion of the spherical surface that it encloses and also the boundary of the complementary region. However, a meridian or parallel circle on the torus is not the boundary of two regions of the torus since such a circle does not separate the torus. Thus any simple closed curve on the sphere is homologous to zero, but meridian and parallel circles on the torus are not homologous to zero.

The following intuitive example will make more precise this still vague idea of homology. It is based on the modulo 2 homology theory introduced by Heinrich Tietze in 1908. Consider the configuration shown in Figure 1.5 consisting of triangles $\langle abc\rangle$, $\langle bcd\rangle$, $\langle abd\rangle$, and $\langle acd\rangle$, edges $\langle ab\rangle$, $\langle ac\rangle$, $\langle ad\rangle$, $\langle bc\rangle$, $\langle bd\rangle$, $\langle cd\rangle$, $\langle df\rangle$, $\langle de\rangle$, $\langle ef\rangle$, and $\langle fg\rangle$, and vertices $\langle a\rangle$, $\langle b\rangle$, $\langle c\rangle$, $\langle d\rangle$, $\langle e\rangle$, $\langle f\rangle$, and $\langle g\rangle$. The interior of the tetrahedron and the interior of triangle $\langle def\rangle$ are not included. This type of space is called a "polyhedron"; the definition of this term will be given in the next section.

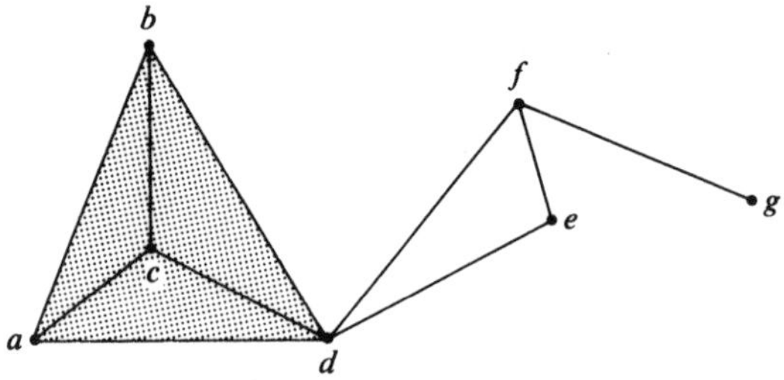

Figure 1.5

A 2-*chain* is a formal linear combination of triangles with coefficients modulo 2. A 1-*chain* is a formal linear combination of edges with coefficients modulo 2. The 0-*chains* are similarly defined for vertices. To simplify the notation, we omit those terms with coefficient 0 and consider only those terms in a chain with coefficient 1. Thus we write

$$\langle abc\rangle + \langle abd\rangle$$

to denote the 2-chain

$$1\cdot\langle abc\rangle + 1\cdot\langle abd\rangle + 0\cdot\langle acd\rangle + 0\cdot\langle bcd\rangle.$$

The *boundary operator* ∂ is defined as follows for chains of length one and extended linearly:

$$\partial\langle abc\rangle = \langle ab\rangle + \langle ac\rangle + \langle bc\rangle,$$
$$\partial\langle ab\rangle = \langle a\rangle + \langle b\rangle.$$

A p-chain c_p ($p = 1$ or 2) is a *boundary* means that there is a $(p + 1)$-chain c_{p+1} with

$$\partial c_{p+1} = c_p.$$

We think of this intuitively as indicating that the union of the members of c_p forms the point-set boundary of the union of the members of c_{p+1}. For example,

$$\langle ab\rangle + \langle bc\rangle + \langle cd\rangle + \langle da\rangle = \partial(\langle abc\rangle + \langle acd\rangle),$$

since terms which occur twice cancel modulo 2. For any 2-chain c_2, one easily observes that

$$\partial\partial c_2 = 0.$$

A *p-cycle* ($p = 1$ or 2) is a p-chain c_p with $\partial c_p = 0$. Since $\partial\partial$ is the trivial operator, then every boundary is a cycle. Intuitively speaking, a cycle is a chain whose terms either close a "hole" or form the boundary of a chain of the next higher dimension. We investigate the "holes" in the polyhedron by determining the cycles which are not boundaries.

Except for the 2-chain having all coefficients zero,

$$\langle abc\rangle + \langle bcd\rangle + \langle acd\rangle + \langle abd\rangle$$

is the only 2-cycle in our example, and it is nonbounding since the interior of the tetrahedron is not included. The reader should check to see that

$$z = \langle df\rangle + \langle fe\rangle + \langle de\rangle$$

is a nonbounding 1-cycle and that any other 1-cycle is either a boundary or the sum of z and a boundary. Thus any 1-cycle is homologous to zero or homologous to the fundamental 1-cycle z. This indicates the presence of two holes in the polyhedron, one enclosed by the nonbounding 2-cycle and one enclosed by the nonbounding 1-cycle z.

In Chapter 2 we shall make rigorous the notions of homology, chain, cycle, and boundary and use them to study the structure of general polyhedra.

1.3 Geometric Complexes and Polyhedra

We turn now to the problem of defining polyhedra, the subspaces of Euclidean n-space $\mathbb{R}^n$ on which homology theory will be developed. Intuitively, a polyhedron is a subset of $\mathbb{R}^n$ composed of vertices, line segments, triangles, tetrahedra, and so on joined together as in the example of mod 2 homology in the preceding section. Naturally we must allow for higher dimensions and considerable generality in the definition.

For each positive integer n, we shall consider n-dimensional Euclidean space

$$\mathbb{R}^n = \{x = (x_1, x_2, \ldots, x_n)\text{: each } x_i \text{ is a real number}\}$$

as a vector space over the field $\mathbb{R}$ of real numbers and use some basic ideas from the theory of vector spaces. The reader who has not studied vector spaces should consult Appendix 3 before proceeding.

Definition. A set $A = \{a_0, a_1, \ldots, a_k\}$ of $k + 1$ points in $\mathbb{R}^n$ is *geometrically independent* means that no hyperplane of dimension $k - 1$ contains all the points.

Thus a set $\{a_0, a_1, \ldots, a_k\}$ is geometrically independent means that all the points are distinct, no three of them lie on a line, no four of them lie in a plane, and, in general, no $p + 1$ of them lie in a hyperplane of dimension $p - 1$ or less.

Example 1.1. The set $\{a_0, a_1, a_2\}$ in Figure 1.6(a) is geometrically independent since the only hyperplane in $\mathbb{R}^2$ containing all the points is the entire plane. The set $\{b_0, b_1, b_2\}$ in Figure 1.6(b) is not geometrically independent since all three points lie on a line, a hyperplane of dimension 1.

Definition. Let $\{a_0, \ldots, a_k\}$ be a set of geometrically independent points in $\mathbb{R}^n$. The *k-dimensional geometric simplex* or *k-simplex*, σ^k, spanned by

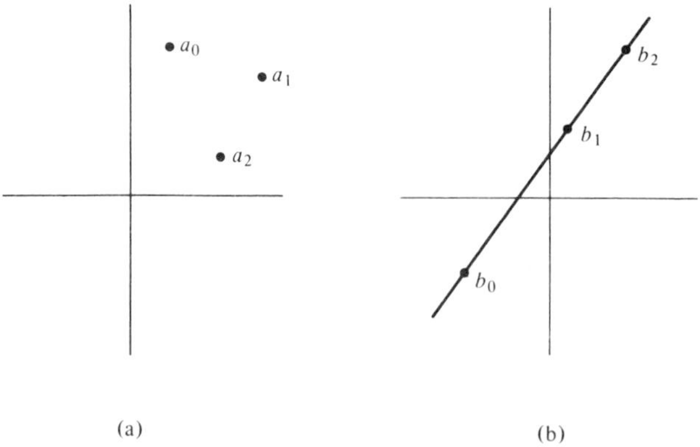

(a) (b)

Figure 1.6

$\{a_0, \ldots, a_k\}$ is the set of all points x in $\mathbb{R}^n$ for which there exist nonnegative real numbers $\lambda_0, \ldots, \lambda_k$ such that

$$x = \sum_{i=0}^{k} \lambda_i a_i, \qquad \sum_{i=0}^{k} \lambda_i = 1.$$

The numbers $\lambda_0, \ldots, \lambda_k$ are the *barycentric coordinates* of the point x. The points $a_0, \ldots, a_k$ are the *vertices* of σ^k. The set of all points x in σ^k with all barycentric coordinates positive is called the *open geometric k-simplex* spanned by $\{a_0, \ldots, a_k\}$.

Observe that a 0-simplex is simply a singleton set, a 1-simplex is a closed line segment, a 2-simplex is a triangle (interior and boundary), and a 3-simplex is a tetrahedron (interior and boundary). An open 0-simplex is a singleton set, an open 1-simplex is a line segment with end points removed, an open 2-simplex is the interior of a triangle, and an open 3-simplex is the interior of a tetrahedron.

Definition. A simplex σ^k is a *face* of a simplex σ^n, $k \leq n$, means that each vertex of σ^k is a vertex of σ^n. The faces of σ^n other than σ^n itself are called *proper faces.*

If σ^n is the simplex with vertices $a_0, \ldots, a_n$, we shall write

$$\sigma^n = \langle a_0 \ldots a_n \rangle.$$

Then the faces of the 2-simplex $\langle a_0 a_1 a_2 \rangle$ are the 2-simplex itself, the 1-simplexes $\langle a_0 a_1 \rangle$, $\langle a_1 a_2 \rangle$, and $\langle a_0 a_2 \rangle$, and the 0-simplexes $\langle a_0 \rangle$, $\langle a_1 \rangle$, and $\langle a_2 \rangle$.

Definition. Two simplexes σ^m and σ^n are *properly joined* provided that they do not intersect or the intersection $\sigma^m \cap \sigma^n$ is a face of both σ^m and σ^n.

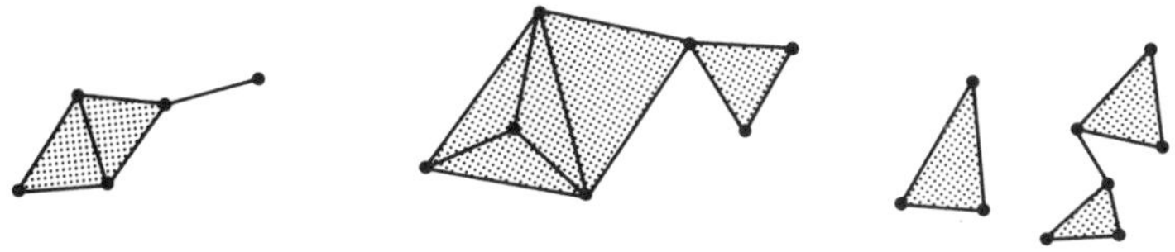

(a) (b) (c)

Figure 1.7 Examples of proper joining

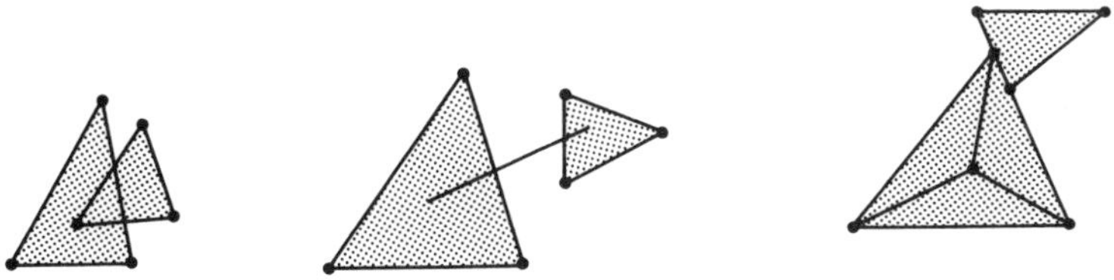

(a) (b) (c)

Figure 1.8 Examples of improper joining

Definition. A *geometric complex* (or *simplicial complex* or *complex*) is a finite family K of geometric simplexes which are properly joined and have the property that each face of a member of K is also a member of K. The *dimension* of K is the largest positive integer r such that K has an r-simplex. The union of the members of K with the Euclidean subspace topology is denoted by $|K|$ and is called the *geometric carrier* of K or the *polyhedron associated with* K.

We shall be concerned, for the purposes of homology, with geometric complexes and polyhedra composed of a finite number of simplexes as defined above. Greater generality, at the expense of greater complexity, can be obtained by allowing an infinite number of simplexes. The reader interested in this generalization should consult the text by Hocking and Young [9].

There are several reasons for restricting our initial considerations to polyhedra. They are easily visualized and are sufficiently general to allow meaningful applications. Poincaré realized this and gave a definition of complex in his second paper on algebraic topology, *Complément à l'Analysis Situs* [50], in 1899. Furthermore, polyhedra are more general than they appear at first glance. A theorem of P. S. Alexandroff (1928) insures that every compact metric space can be indefinitely approximated by polyhedra. This allows us to carry over some topological theorems about polyhedra to compacta by suitable limiting processes. After a thorough introduction to homology theory of polyhedra, we shall look at one of its generalizations, singular homology theory, which applies to all topological spaces.

Definition. Let X be a topological space. If there is a geometric complex K whose geometric carrier $|K|$ is homeomorphic to X, then X is said to be a *triangulable space*, and the complex K is called a *triangulation* of X.

Definition. The *closure* of a k-simplex σ^k, $\mathrm{Cl}(\sigma^k)$, is the complex consisting of σ^k and all its faces.

Definition. If K is a complex and r a positive integer, the *r-skeleton* of K is the complex consisting of all simplexes of K of dimension less than or equal to r.

Example 1.2. (a) Consider a 3-simplex $\sigma^3 = \langle a_0a_1a_2a_3\rangle$. The 2-skeleton of the closure of σ^3 is the complex K whose simplexes are the proper faces of σ^3. The geometric carrier of K is the boundary of a tetrahedron and is therefore homeomorphic to the 2-sphere

$$S^2 = \left\{(x_1, x_2, x_3) \in \mathbb{R}^3 : \sum_{i=1}^{3} x_i^2 = 1\right\}.$$

Thus S^2 is triangulable with K as one triangulation.

(b) The n-sphere

$$S^n = \left\{(x_1, x_2, \ldots, x_{n+1}) \in \mathbb{R}^{n+1} : \sum_{i=1}^{n+1} x_i^2 = 1\right\}$$

is a triangulable space for $n \geq 0$. The n-skeleton of the closure of an $(n + 1)$-simplex σ^{n+1} is one triangulation of S^n. The reader should verify this by solving Exercise 12.

(c) The *Möbius strip* is obtained by identifying two opposite ends of a rectangle after twisting it through 180 degrees. This can easily be done with a strip of paper. Figure 1.9 shows a triangulation of the Möbius strip. It is understood that the two vertices labeled a_0 are identified, the two vertices labeled a_3 are identified, corresponding points of the two segments $\langle a_0a_3\rangle$ are identified, and the resulting quotient space, the geometric carrier of the triangulation, is considered as a subspace of $\mathbb{R}^3$.

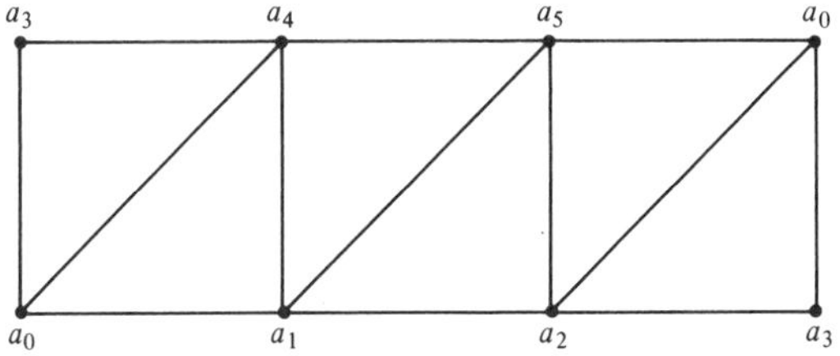

Figure 1.9

(d) A *torus* is obtained from a cylinder by identifying corresponding points of the circular ends with no twisting, as shown in Figure 1.10.

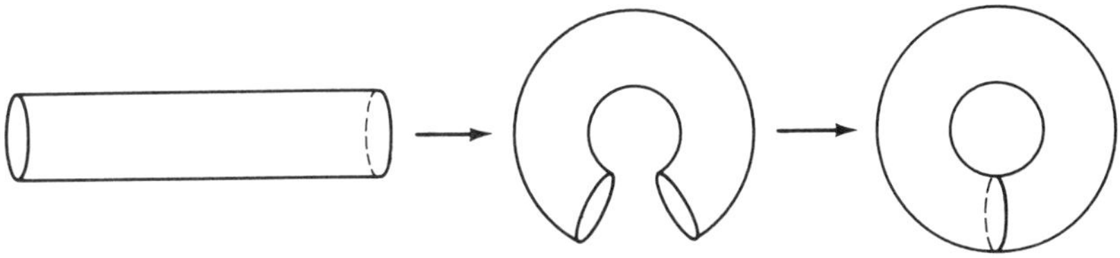

Figure 1.10

Verify the fact that the following diagram, with proper identifications, gives a triangulation of the torus.

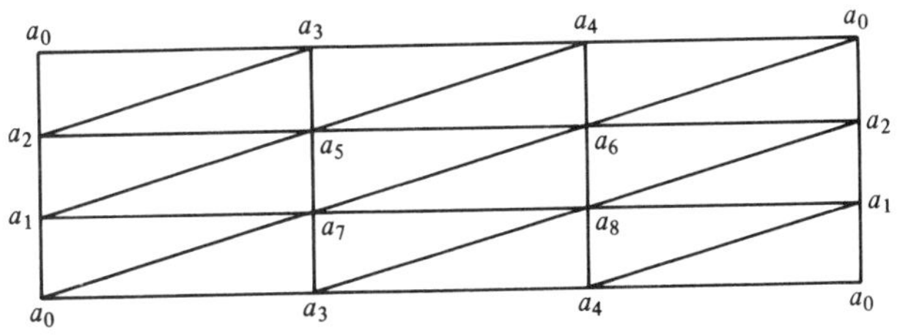

Figure 1.11

1.4 Orientation of Geometric Complexes

Definition. An *oriented n-simplex*, $n \geq 1$, is obtained from an n-simplex $\sigma^n = \langle a_0 \ldots a_n \rangle$ by choosing an ordering for its vertices. The equivalence class of even permutations of the chosen ordering determines the *positively oriented simplex* $+\sigma^n$ while the equivalence class of odd permutations determines the *negatively oriented simplex* $-\sigma^n$. An *oriented geometric complex* is obtained from a geometric complex by assigning an orientation to each of its simplexes.

If vertices $a_0, \ldots, a_p$ of a complex K are the vertices of a p-simplex σ^p, then the symbol $+\langle a_0 \ldots a_p \rangle$ denotes the class of even permutations of the indicated order $a_0, \ldots, a_p$ and $-\langle a_0 \ldots a_p \rangle$ denotes the class of odd permutations. If we wanted the class of even permutations of this order to determine the positively oriented simplex, then we would write

$$+\sigma^p = \langle a_0 \ldots a_p \rangle$$

or

$$+\sigma^p = +\langle a_0 \ldots a_p \rangle.$$

Since ordering vertices requires more than one vertex, we need not worry about orienting 0-simplexes. It will be convenient, however, to consider a 0-simplex $\langle a_0 \rangle$ as positively oriented.

Example 1.3. (a) In the 1-simplex $\sigma^1 = \langle a_0 a_1 \rangle$, let us agree that the ordering is given by $a_0 < a_1$. Then

$$+\sigma^1 = \langle a_0 a_1 \rangle, \qquad -\sigma^1 = \langle a_1 a_0 \rangle.$$

If we imagine that the segment $\langle a_i a_j \rangle$ is directed from a_i toward a_j, then $\langle a_0 a_1 \rangle$ and $\langle a_1 a_0 \rangle$ have opposite directions.

(b) In the 2-simplex $\sigma^2 = \langle a_0 a_1 a_2 \rangle$, assign the order $a_0 < a_1 < a_2$. Then $\langle a_0 a_1 a_2 \rangle$, $\langle a_1 a_2 a_0 \rangle$, and $\langle a_2 a_0 a_1 \rangle$ all denote $+\sigma^2$, while $\langle a_0 a_2 a_1 \rangle$, $\langle a_2 a_1 a_0 \rangle$, and $\langle a_1 a_0 a_2 \rangle$ all denote $-\sigma^2$. (See Figure 1.12.) Then

$$+\sigma^2 = +\langle a_0 a_1 a_2 \rangle, \qquad -\sigma^2 = -\langle a_0 a_1 a_2 \rangle = +\langle a_0 a_2 a_1 \rangle.$$

(Here $+\langle a_0 a_2 a_1 \rangle$ denotes the class of even permutations of a_0, a_2, a_1, and $-\langle a_0 a_1 a_2 \rangle$ denotes the class of odd permutations of a_0, a_1, and a_2.)

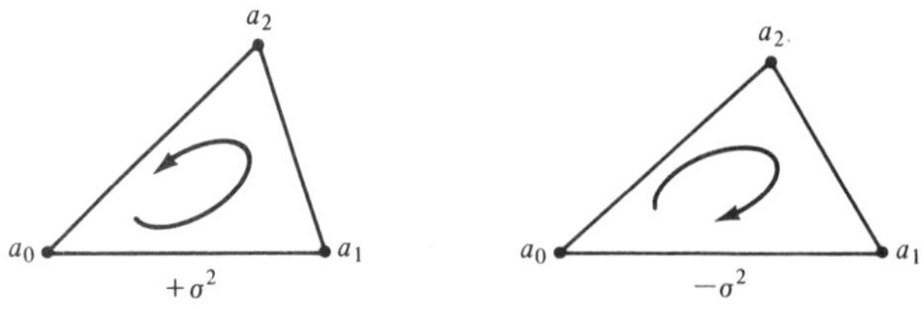

Figure 1.12

One method of orienting a complex is to choose an ordering for all its vertices and to use this ordering to induce an ordering on the vertices of each simplex. This is not the only method, however. An orientation may be assigned to each simplex individually without regard to the manner in which the simplexes are joined. From this point on, we assume that each complex under consideration is assigned some orientation.

Here is a word of comfort for those who suspect that different orientations will introduce great complexity into our considerations: they won't. We are developing a method of describing the topological structure of a polyhedron $|K|$ by determining the "holes" and "twisting" which occur in the associated complex K. In the final analysis, the determining factor is the topological structure of $|K|$ and not the particular triangulation nor the particular orientation. A triangulation is a convenient method of visualizing the polyhedron and converting it to a standard form. An orientation is simply a convenient vehicle for cataloguing the arrangement of the simplexes. Neither the particular triangulation nor the particular orientation makes any difference in the final outcome.

Definition. Let K be an oriented geometric complex with simplexes σ^{p+1} and σ^p whose dimensions differ by 1. We associate with each such pair (σ^{p+1}, σ^p) an *incidence number* $[\sigma^{p+1}, \sigma^p]$ defined as follows: If σ^p is not a face of σ^{p+1}, then $[\sigma^{p+1}, \sigma^p] = 0$. Suppose σ^p is a face of σ^{p+1}. Label the vertices $a_0, \ldots, a_p$ of σ^p so that $+\sigma^p = +\langle a_0 \ldots a_p\rangle$. Let v denote the vertex of σ^{p+1} which is not in σ^p. Then $+\sigma^{p+1} = \pm\langle va_0 \ldots a_p\rangle$. If $+\sigma^{p+1} = +\langle va_0 \ldots a_p\rangle$, then $[\sigma^{p+1}, \sigma^p] = 1$. If $+\sigma^{p+1} = -\langle va_0 \ldots a_p\rangle$, then $[\sigma^{p+1}, \sigma^p] = -1$.

Example 1.4. (a) If $+\sigma^1 = \langle a_0a_1\rangle$, then $[\sigma^1, \langle a_0\rangle] = -1$ and $[\sigma^1, \langle a_1\rangle] = 1$.

(b) If $+\sigma^2 = +\langle a_0a_1a_2\rangle$, $+\sigma^1 = \langle a_0a_1\rangle$ and $+\tau^1 = \langle a_0a_2\rangle$, then $[\sigma^2, \sigma^1] = 1$ and $[\sigma^2, \tau^1] = -1$.

Note that in Figure 1.12 the arrow indicating the orientation of σ^2 agrees with the orientation of σ^1 but disagrees with the orientation of τ^1.

Theorem 1.1. *Let K be an oriented complex, σ^p an oriented p-simplex of K and σ^{p-2} a $(p-2)$-face of σ^p. Then*

$$\sum [\sigma^p, \sigma^{p-1}][\sigma^{p-1}, \sigma^{p-2}] = 0, \qquad \sigma^{p-1} \in K.$$

Proof. Label the vertices $v_0, \ldots, v_{p-2}$ of σ^{p-2} so that $+\sigma^{p-2} = \langle v_0 \ldots v_{p-2}\rangle$. Then σ^p has two additional vertices a and b, and we may assume that $+\sigma^p = \langle abv_0 \ldots v_{p-2}\rangle$. Nonzero terms occur in the sum for only two values of σ^{p-1}, namely

$$\sigma_1^{p-1} = \langle av_0 \ldots v_{p-2}\rangle, \qquad \sigma_2^{p-1} = \langle bv_0 \ldots v_{p-2}\rangle.$$

We must now treat four cases determined by the orientations of σ_1^{p-1} and σ_2^{p-1}.

Case I. Suppose that

$$+\sigma_1^{p-1} = +\langle av_0 \ldots v_{p-2}\rangle, \qquad +\sigma_2^{p-1} = +\langle bv_0 \ldots v_{p-2}\rangle.$$

Then

$$[\sigma^p, \sigma_1^{p-1}] = -1, \qquad [\sigma_1^{p-1}, \sigma^{p-2}] = +1,$$
$$[\sigma^p, \sigma_2^{p-1}] = +1, \qquad [\sigma_2^{p-1}, \sigma^{p-2}] = +1,$$

so that the sum of the indicated products is 0.

Case II. Suppose that

$$+\sigma_1^{p-1} = +\langle av_0 \ldots v_{p-2}\rangle, \qquad +\sigma_2^{p-1} = -\langle bv_0 \ldots v_{p-2}\rangle.$$

Then

$$[\sigma^p, \sigma_1^{p-1}] = -1, \qquad [\sigma_1^{p-1}, \sigma^{p-2}] = +1,$$
$$[\sigma^p, \sigma_2^{p-1}] = -1, \qquad [\sigma_2^{p-1}, \sigma^{p-2}] = -1,$$

so that the desired conclusion holds in this case also.

The remaining cases are left as an exercise. □

Definition. In the oriented complex K, let $\{\sigma_i^p\}_{i=1}^{\alpha_p}$ and $\{\sigma_i^{p+1}\}_{i=1}^{\alpha_{p+1}}$ denote the p-simplexes and $(p + 1)$-simplexes of K, where α_p and α_{p+1} denote the numbers of simplexes of dimensions p and $p + 1$ respectively. The matrix

$$\eta(p) = (\eta_{ij}(p)),$$

where $\eta_{ij}(p) = [\sigma_i^{p+1}, \sigma_j^p]$, is called the *pth incidence matrix* of K.

Incidence matrices were used to describe the arrangement of simplexes in a complex during the early days of algebraic or "combinatorial" topology. They are less in vogue today because group theory has given a much more efficient method of describing the same property. The group theoretic formulation seems to have been suggested by the famous algebraist Emmy Noether (1882–1935) about 1925. As we shall see in Chapter 2, these groups follow quite naturally from Poincaré's original description of homology theory.

Exercises

1. Fill in the details of the mod 2 homology example given in the text.

2. Prove that a set of $k + 1$ points in $\mathbb{R}^n$ is geometrically independent if and only if no $p + 1$ of the points lie in a hyperplane of dimension less than or equal to $p - 1$.

3. Prove that a set $A = \{a_0, a_1, \ldots, a_k\}$ of points in $\mathbb{R}^n$ is geometrically independent if and only if the set of vectors $\{a_1 - a_0, \ldots, a_k - a_0\}$ is linearly independent.

4. Show that the barycentric coordinates of each point in a simplex are unique.

5. A subset B of $\mathbb{R}^n$ is *convex* provided that B contains every line segment having two of its members as end points.

(a) If a and b are points in $\mathbb{R}^n$, show that the line segment L joining a and b consists of all points x of the form

$$x = ta + (1 - t)b$$

where t is a real number with $0 \leq t \leq 1$.

(b) Show that every simplex is a convex set.

(c) Prove that a simplex σ is the smallest convex set which contains all vertices of σ.

6. How many faces does an n-simplex have? Prove that your answer is correct.

7. Verify that the r-skeleton of a geometric complex is a geometric complex.

8. The *Klein Bottle* is obtained from a cylinder by identifying the two circular ends with the orientation of the two circles reversed. (It cannot be constructed in 3-dimensional space without self-intersection.) Modify the triangulation of the torus given in the text to produce a triangulation of the Klein Bottle.

9. Let K denote the closure of a 3-simplex $\sigma^3 = \langle a_0 a_1 a_2 a_3 \rangle$ with vertices ordered by

$$a_0 < a_1 < a_2 < a_3.$$

Use this given order to induce an orientation on each simplex of K, and determine all incidence numbers associated with K.

10. Complete the proof of Theorem 1.

11. In the triangulation M of the Möbius strip in Figure 1.9, let us call a 1-simplex *interior* if it is a face of two 2-simplexes. For each interior simplex σ_i, let $\bar{\sigma}_i$ and $\bar{\bar{\sigma}}_i$ denote the two 2-simplexes of which σ_i is a face. Show that it is not possible to orient M so that

$$[\bar{\sigma}_i, \sigma_i] = -[\bar{\bar{\sigma}}_i, \sigma_i]$$

for each interior simplex σ_i. (This result is sometimes expressed by saying that M is *nonorientable* or that it has no *coherent orientation*.)

12. Let $\sigma^{n+1} = \langle a_0 \ldots a_{n+1} \rangle$ be the $(n + 1)$-simplex in $\mathbb{R}^{n+1}$ with vertices as follows: a_0 is the origin and, for $i \geq 1$, a_i is the point with ith coordinate 1 and all other coordinates 0. Let K denote the n-skeleton of the closure of σ^{n+1}. Show that S^n is triangulable by exhibiting a homeomorphism between S^n and $|K|$. (Hint: If σ^{n+1} is considered as a subspace of $\mathbb{R}^{n+1}$, then $|K|$ is its point-set boundary.)

2 Simplicial Homology Groups

Having defined polyhedron, complex, and orientation for complexes in the preceding chapter, we are now ready for the precise definition of the homology groups. Intuitively speaking, the homology groups of a complex describe the arrangement of the simplexes in the complex thereby telling us about the "holes" in the associated polyhedron.

Whether expressly stated or not, we assume that each complex under consideration has been assigned an orientation.

2.1 Chains, Cycles, Boundaries, and Homology Groups

Definition. Let K be an oriented simplicial complex. If p is a positive integer, a *p-dimensional chain*, or *p-chain*, is a function c_p from the family of oriented p-simplexes of K to the integers such that, for each p-simplex σ^p, $c_p(-\sigma^p) = -c_p(+\sigma^p)$. A *0-dimensional chain* or 0-chain is a function from the 0-simplexes of K to the integers. With the operation of pointwise addition induced by the integers, the family of p-chains forms a group called the *p-dimensional chain group* of K. This group is denoted by $C_p(K)$.

An *elementary p-chain* is a p-chain c_p for which there is a p-simplex σ^p such that $c_p(\tau^p) = 0$ for each p-simplex τ^p distinct from σ^p. Such an elementary p-chain is denoted by $g \cdot \sigma^p$ where $g = c_p(+\sigma^p)$. With this notation, an arbitrary p-chain d_p can be expressed as a formal finite sum

$$d_p = \sum g_i \cdot \sigma_i^p$$

of elementary p-chains where the index i ranges over all p-simplexes of K.

The following facts should be observed from the definition of p-chains:

(a) If $c_p = \sum f_i \cdot \sigma_i^p$ and $d_p = \sum g_i \cdot \sigma_i^p$ are two p-chains on K, then

$$c_p + d_p = \sum (f_i + g_i) \cdot \sigma_i^p.$$

(b) The additive inverse of the chain c_p in the group $C_p(K)$ is the chain $-c_p = \sum -f_i \cdot \sigma_i^p$.

(c) The chain group $C_p(K)$ is isomorphic to the direct sum of the group $\mathbb{Z}$ of integers over the family of p-simplexes of K. That is, if K has α_p p-simplexes, then $C_p(K)$ is isomorphic to the direct sum of α_p copies of $\mathbb{Z}$. One isomorphism is given by the correspondence

$$\sum_{i=1}^{\alpha_p} g_i \cdot \sigma_i^p \leftrightarrow (g_1, g_2, \ldots, g_{\alpha_p}).$$

Algebraic systems other than the integers could be used as the coefficient set for the p-chains. Any commutative group, commutative ring, or field could be used thus making $C_p(K)$ a commutative group, a module, or a vector space. With two exceptions, we shall use only the integers as the coefficient set for chains. Incidentally, Poincaré's original definition was given in terms of integers.

Definition. If $g \cdot \sigma^p$ is an elementary p-chain with $p \geq 1$, the *boundary* of $g \cdot \sigma^p$, denoted by $\partial(g \cdot \sigma^p)$, is defined by

$$\partial(g \cdot \sigma^p) = \sum [\sigma^p, \sigma_i^{p-1}] g \cdot \sigma_i^{p-1}, \qquad \sigma_i^{p-1} \in K.$$

The boundary operator ∂ is extended by linearity to a homomorphism

$$\partial\colon C_p(K) \to C_{p-1}(K).$$

In other words, if $c_p = \sum g_i \cdot \sigma_i^p$ is an arbitrary p-chain, then we define

$$\partial(c_p) = \sum \partial(g_i \cdot \sigma_i^p).$$

The *boundary of a* 0-*chain* is defined to be zero.

Strictly speaking, we should say that there is a boundary homomorphism

$$\partial_p\colon C_p(K) \to C_{p-1}(K).$$

This extra subscript is cumbersome, however, and we shall usually omit it since the dimension involved is indicated by the chain group $C_p(K)$.

Theorem 2.1. *If K is an oriented complex and $p \geq 2$, then the composition $\partial\partial\colon C_p(K) \to C_{p-2}(K)$ in the diagram*

$$C_p(K) \xrightarrow{\partial} C_{p-1}(K) \xrightarrow{\partial} C_{p-2}(K)$$

is the trivial homomorphism.

Proof. We must prove that $\partial\partial(c_p) = 0$ for each p-chain. To do this, it is sufficient to show that $\partial\partial(g \cdot \sigma^p) = 0$ for each elementary p-chain $g \cdot \sigma^p$.

Observe that

$$\partial\partial(g \cdot \sigma^p) = \partial\left(\sum_{\sigma_i^{p-1} \in K} [\sigma^p, \sigma_i^{p-1}] g \cdot \sigma_i^{p-1}\right) = \sum_{\sigma_i^{p-1} \in K} \partial([\sigma^p, \sigma_i^{p-1}] g \cdot \sigma_i^{p-1})$$

$$= \sum_{\sigma_i^{p-1} \in K} \sum_{\sigma_j^{p-2} \in K} [\sigma^p, \sigma_i^{p-1}][\sigma_i^{p-1}, \sigma_j^{p-2}] g \cdot \sigma_j^{p-2}.$$

Reversing the order of summation and collecting coefficients of each simplex σ_j^{p-2} gives

$$\partial\partial(g\cdot\sigma^p) = \sum_{\sigma_j^{p-2}\in K}\left(\sum_{\sigma_i^{p-1}\in K}[\sigma^p, \sigma_i^{p-1}][\sigma_i^{p-1}, \sigma_j^{p-2}]g\cdot\sigma_j^{p-2}\right).$$

Since Theorem 1.1 insures that $\sum_{\sigma_i^{p-1}\in K}[\sigma^p, \sigma_i^{p-1}][\sigma_i^{p-1}, \sigma_j^{p-2}]$ is 0 for each σ_j^{p-2}, it follows that $\partial\partial(g\cdot\sigma^p) = 0$. □

Definition. Let K be an oriented complex. If p is a positive integer, a *p-dimensional cycle* on K, or *p-cycle*, is a p-chain z_p such that $\partial(z_p) = 0$. The family of p-cycles is thus the kernel of the homomorphism $\partial: C_p(K) \to C_{p-1}(K)$ and is a subgroup of $C_p(K)$. This subgroup, denoted by $Z_p(K)$, is called the *p-dimensional cycle group* of K. Since we have defined the boundary of every 0-chain to be 0, we now define *0-cycle* to be synonymous with 0-chain. Thus the group $Z_0(K)$ of 0-cycles is the group $C_0(K)$ of 0-chains.

If $p \geq 0$, a p-chain b_p is a *p-dimensional boundary* on K, or *p-boundary*, if there is a $(p + 1)$-chain c_{p+1} such that $\partial(c_{p+1}) = b_p$. The family of p-boundaries is the homomorphic image $\partial(C_{p+1}(K))$ and is a subgroup of $C_p(K)$. This subgroup is called the *p-dimensional boundary group* of K and is denoted by $B_p(K)$.

If n is the dimension of K, then there are no p-chains on K for $p > n$. In this case we say that $C_p(K)$ is the trivial group $\{0\}$. In particular, there are no $(n + 1)$-chains on K so that $C_{n+1}(K) = \{0\}$ and therefore $B_n(K) = \{0\}$.

The proof of the following theorem is left as an exercise:

Theorem 2.2. *If K is an oriented complex, then $B_p(K) \subset Z_p(K)$ for each integer p such that $0 \leq p \leq n$, where n is the dimension of K.*

We think intuitively of a p-cycle as a linear combination of p-simplexes which makes a complete circuit. The p-cycles which enclose "holes" are the interesting cycles, and they are the ones which are not boundaries of $(p + 1)$-chains. We restrict our attention to nonbounding cycles and weed out the bounding ones. A p-cycle which is the boundary of a $(p + 1)$-chain was said by Poincaré to be *homologous to zero.* The separation of cycles into these categories is accomplished by the following definition.

Definition. Two p-cycles w_p and z_p on a complex K are *homologous,* written $w_p \sim z_p$, provided that there is a $(p + 1)$-chain c_{p+1} such that

$$\partial(c_{p+1}) = w_p - z_p.$$

If a p-cycle t_p is the boundary of a $(p + 1)$-chain, we say that t_p is *homologous to zero* and write $t_p \sim 0$.

This relation of homology for p-cycles is an equivalence relation and partitions $Z_p(K)$ into *homology classes*

$$[z_p] = \{w_p \in Z_p(K) : w_p \sim z_p\}.$$

The homology class $[z_p]$ is actually the coset

$$z_p + B_p(K) = \{z_p + \partial(c_{p+1}) : \partial(c_{p+1}) \in B_p(K)\}.$$

Hence the homology classes are actually the members of the quotient group $Z_p(K)/B_p(K)$. We can use the quotient group structure to add homology classes.

Definition. If K is an oriented complex and p a non-negative integer, the *p-dimensional homology group* of K is the quotient group

$$H_p(K) = Z_p(K)/B_p(K).$$

2.2 Examples of Homology Groups

The following examples are intended to clarify the preceding definitions:

Example 2.1. Let K be the closure of a 2-simplex $\langle a_0a_1a_2\rangle$ with orientation induced by the ordering $a_0 < a_1 < a_2$. Thus K has 0-simplexes $\langle a_0\rangle$, $\langle a_1\rangle$, and $\langle a_2\rangle$, positively oriented 1-simplexes $\langle a_0a_1\rangle$, $\langle a_1a_2\rangle$, and $\langle a_0a_2\rangle$ and positively oriented 2-simplex $\langle a_0a_1a_2\rangle$.

A 0-chain on K is a sum of the form

$$c_0 = g_0\cdot\langle a_0\rangle + g_1\cdot\langle a_1\rangle + g_2\cdot\langle a_2\rangle$$

where g_0, g_1, and g_2 are integers. Hence $C_0(K) = Z_0(K)$ is isomorphic to the direct sum $\mathbb{Z} \oplus \mathbb{Z} \oplus \mathbb{Z}$ of three copies of the group of integers. A 1-chain on K is a sum of the form

$$c_1 = h_0\cdot\langle a_0a_1\rangle + h_1\cdot\langle a_1a_2\rangle + h_2\cdot\langle a_0a_2\rangle$$

where h_0, h_1, and h_2 are integers, so $C_1(K)$ is isomorphic to $\mathbb{Z} \oplus \mathbb{Z} \oplus \mathbb{Z}$. Also,

$$\partial(c_1) = (-h_0 - h_2)\cdot\langle a_0\rangle + (h_0 - h_1)\cdot\langle a_1\rangle + (h_1 + h_2)\cdot\langle a_2\rangle. \tag{1}$$

Hence c_1 is a 1-cycle if and only if h_0, h_1, and h_2 satisfy the equations

$$-h_0 - h_2 = 0, \qquad h_0 - h_1 = 0, \qquad h_1 + h_2 = 0.$$

This system gives $h_0 = h_1 = -h_2$ so that the 1-cycles are chains of the form

$$h\cdot\langle a_0a_1\rangle + h\cdot\langle a_1a_2\rangle - h\cdot\langle a_0a_2\rangle \tag{2}$$

where h is any integer. Thus $Z_1(K)$ is isomorphic to the group $\mathbb{Z}$ of integers.

The only 2-simplex of K is $\langle a_0a_1a_2\rangle$, so the only 2-chains are the elementary ones $h\cdot\langle a_0a_1a_2\rangle$ where h is an integer. Thus $C_2(K) \cong \mathbb{Z}$. Since

$$\partial(h\cdot\langle a_0a_1a_2\rangle) = h\cdot\langle a_0a_1\rangle + h\cdot\langle a_1a_2\rangle - h\cdot\langle a_0a_2\rangle, \tag{3}$$

then $\partial(h\cdot\langle a_0a_1a_2\rangle) = 0$ only when $h = 0$. Thus $Z_2(K) = \{0\}$, so $H_2(K) = \{0\}$.

From Equations (2) and (3), we observe that 1-cycles and 1-boundaries have precisely the same form so that $Z_1(K) = B_1(K)$, and hence $H_1(K) = \{0\}$.

From Equation (1) we observe that a 0-cycle

$$g_0\cdot\langle a_0\rangle + g_1\cdot\langle a_1\rangle + g_2\cdot\langle a_2\rangle \tag{4}$$

is a 0-boundary if and only if there are integers h_0, h_1, and h_2 such that

$$-h_0 - h_2 = g_0, \qquad h_0 - h_1 = g_1, \qquad h_1 + h_2 = g_2.$$

Then $g_0 + g_1 = -g_2$ so that, for 0-boundaries, two coefficients are arbitrary, and the third is determined by the first two. Thus $B_0(K) \cong \mathbb{Z} \oplus \mathbb{Z}$. Since $Z_0(K) \cong \mathbb{Z} \oplus \mathbb{Z} \oplus \mathbb{Z}$, we now suspect that $H_0(K) \cong \mathbb{Z}$.

To complete the proof, observe that for any 0-cycle expressed in Equation (4),

$$\begin{aligned} &g_0\cdot\langle a_0\rangle + g_1\cdot\langle a_1\rangle + g_2\cdot\langle a_2\rangle \\ &\qquad = \partial(g_1\cdot\langle a_0a_1\rangle + g_2\cdot\langle a_0a_2\rangle) + (g_0 + g_1 + g_2)\cdot\langle a_0\rangle. \end{aligned}$$

This means that any 0-cycle is homologous to a 0-cycle of the form $t\cdot\langle a_0\rangle$, t an integer. Hence each 0-homology class has a representative $t\cdot\langle a_0\rangle$ so that $H_0(K)$ is isomorphic to $\mathbb{Z}$.

Summarizing the above calculations, we have $H_0(K) \cong \mathbb{Z}$, $H_1(K) = \{0\}$, and $H_2(K) = \{0\}$. The trivial groups $H_1(K)$ and $H_2(K)$ indicate the absence of holes in the polyhedron $|K|$. As we shall see later, the fact that $H_0(K)$ is isomorphic to $\mathbb{Z}$ indicates that $|K|$ has one component.

Example 2.2. Let M denote the triangulation of the Möbius strip shown in Figure 2.1 with orientation induced by the ordering $a_0 < a_1 < a_2 < a_3 < a_4 < a_5$.

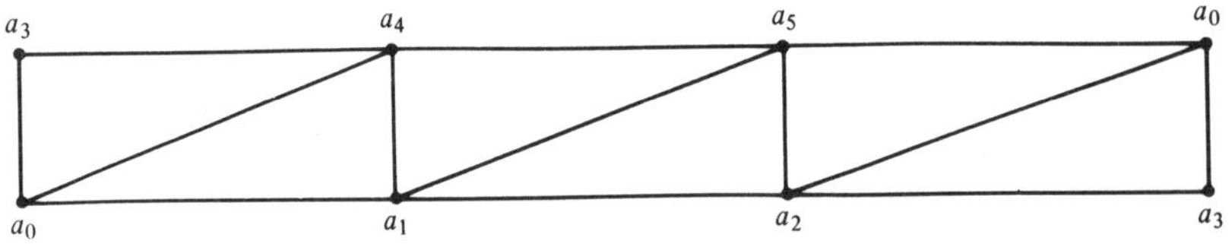

Figure 2.1

There are no 3-simplexes in M, so $B_2(M) = \{0\}$. Suppose that

$$\begin{aligned} w = g_0\cdot\langle a_0a_3a_4\rangle + g_1\cdot\langle a_0a_1a_4\rangle + g_2\cdot\langle a_1a_4a_5\rangle + g_3\cdot\langle a_1a_2a_5\rangle \\ + g_4\cdot\langle a_0a_2a_5\rangle + g_5\cdot\langle a_0a_2a_3\rangle \end{aligned}$$

is a 2-cycle. When $\partial(w)$ is computed, the coefficient that appears with $\langle a_3a_4\rangle$ is g_0. In order to have $\partial(w) = 0$, it must be true that $g_0 = 0$. Similar reasoning applied to the other horizontal 1-simplexes shows that each coefficient in w must be 0. Thus $Z_2(M) = \{0\}$, so $H_2(M) = \{0\}$. Using a bit of intuition, we suspect that the 1-chains

$$\begin{aligned} z &= 1\cdot\langle a_0a_1\rangle + 1\cdot\langle a_1a_2\rangle + 1\cdot\langle a_2a_3\rangle - 1\cdot\langle a_0a_3\rangle, \\ z' &= 1\cdot\langle a_0a_3\rangle + 1\cdot\langle a_3a_4\rangle + 1\cdot\langle a_4a_5\rangle - 1\cdot\langle a_0a_5\rangle \end{aligned}$$

are 1-cycles. (Both of these chains make complete circuits beginning at a_0.) Direct computation verifies that z and z' are cycles. However, $z - z'$ traverses the boundary of M, so $z - z'$ should be the boundary of some 2-chain. A straightforward computation shows that

$$z - z' = \partial(1\cdot\langle a_0a_1a_4\rangle + 1\cdot\langle a_1a_2a_5\rangle + 1\cdot\langle a_0a_2a_3\rangle - 1\cdot\langle a_0a_2a_5\rangle - 1\cdot\langle a_1a_4a_5\rangle - 1\cdot\langle a_0a_3a_4\rangle)$$

so that $z \sim z'$.

A similar calculation verifies the fact that any 1-cycle is homologous to a multiple of z. Hence $H_1(M) = \{[gz]: g \text{ is an integer}\}$, so $H_1(M) \cong \mathbb{Z}$. This result indicates that the polyhedron $|M|$ has one hole bounded by 1-simplexes.

To determine $H_0(M)$, observe that any twoeleme ntary 0-chains $1\cdot\langle a_i\rangle$ and $1\cdot\langle a_j\rangle$ (i, j range from 0 to 5) are homologous. For example,

$$1\cdot\langle a_5\rangle - 1\cdot\langle a_0\rangle = \partial(1\cdot\langle a_0a_4\rangle + 1\cdot\langle a_4a_5\rangle).$$

Hence $H_0(M) = \{[g\cdot\langle a_0\rangle]: g \text{ is an integer}\}$, so $H_0(M) \cong \mathbb{Z}$. As in the preceding example, this indicates that $|M|$ has only one component.

Example 2.3. The *projective plane* is obtained from a finite disk by identifying each pair of diametrically opposite points. A triangulation P of the projective plane, with orientations indicated by the arrows, is shown in Figure 2.2.

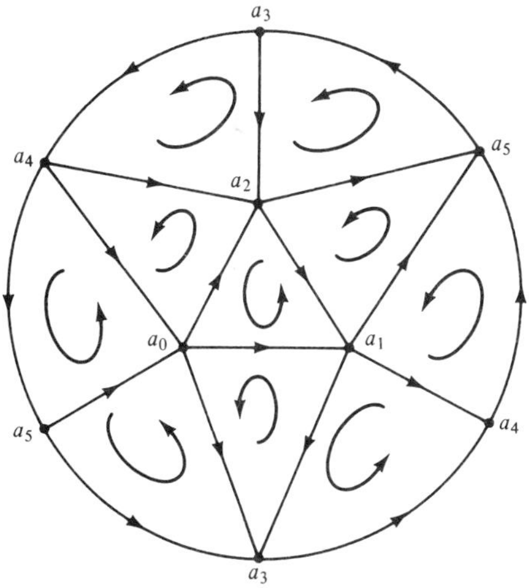

Figure 2.2

There are no 3-simplexes, so $B_2(P) = \{0\}$. To compute $Z_2(P)$, observe that each 1-simplex σ^1 of P is a face of exactly two 2-simplexes σ_1^2 and σ_2^2. Observe that when σ^1 is $\langle a_3a_4\rangle$, $\langle a_4a_5\rangle$, or $\langle a_5a_3\rangle$, both incidence numbers $[\sigma_1^2, \sigma^1]$ and $[\sigma_2^2, \sigma^1]$ are $+1$. For all other choices of σ^1, the two incidence numbers are negatives of each other. Let us call $\langle a_3a_4\rangle$, $\langle a_4a_5\rangle$, and $\langle a_5a_3\rangle$ 1-simplexes of type I and the others 1-simplexes of type II.

Suppose that w is a 2-cycle. In order for the coefficients of the type II 1-simplexes in $\partial(w)$ to be 0, all the coefficients in w must have a common value, say g. But then

$$\partial(w) = 2g\cdot\langle a_3a_4\rangle + 2g\cdot\langle a_4a_5\rangle + 2g\cdot\langle a_5a_3\rangle \tag{5}$$

since both incidence numbers for the type I 1-simplexes are $+1$. Hence w is a 2-cycle only when $g = 0$, so $Z_2(P) = \{0\}$ and $H_2(P) = \{0\}$.

Observe that any 1-cycle is homologous to a multiple of

$$z = 1\cdot\langle a_3a_4\rangle + 1\cdot\langle a_4a_5\rangle + 1\cdot\langle a_5a_3\rangle.$$

Furthermore, Equation (5) shows that any even multiple of z is a boundary. Thus $H_1(P) \cong \mathbb{Z}_2$, the group of integers modulo 2. This result indicates the twisting that occurs around the "hole" in the polyhedron $|P|$. (Recall, however, that the homology groups overlooked the twisted nature of the Möbius strip.)

In the computation of homology groups, it is sometimes convenient to express an elementary chain in terms of a negatively oriented simplex. In order to be able to do this later, let us agree that the symbol $g\cdot(-\sigma^p)$ may be used to denote the elementary p-chain $-g\cdot\sigma^p$. In other words, if $\langle a_0 \ldots a_p\rangle$ represents a positively or negatively oriented p-simplex, then $g\cdot\langle a_0 \ldots a_p\rangle$ denotes the elementary p-chain which assigns value g to the orientation determined by the class of even permutations of the given ordering and assigns value $-g$ to the orientation determined by the class of odd permutations. Return to Example 2.3 for an illustration of this notation. In that example, $\langle a_5a_3\rangle$ denotes a positively oriented 1-simplex. The symbols $g\cdot\langle a_5a_3\rangle$ and $-g\cdot\langle a_3a_5\rangle$ now denote the same elementary 1-chain. An elementary 2-chain $h\cdot\langle a_0a_1a_2\rangle$ may be written in any of six ways:

$$\begin{aligned} h\cdot\langle a_0a_1a_2\rangle &= h\cdot\langle a_1a_2a_0\rangle = h\cdot\langle a_2a_0a_1\rangle = -h\cdot\langle a_1a_0a_2\rangle \\ &= -h\cdot\langle a_0a_2a_1\rangle = -h\cdot\langle a_2a_1a_0\rangle. \end{aligned}$$

2.3 The Structure of Homology Groups

What possibilities are there for the homology groups $H_p(K)$ of a complex K if we take our coefficient group to be the integers? The answer is provided by group theoretic considerations.

Suppose that K has α_p p-simplexes. Then $C_p(K)$ is isomorphic to $\mathbb{Z} \oplus \cdots \oplus \mathbb{Z}$ (α_p summands). In other words, $C_p(K)$ is a free abelian group on α_p generators. Since every subgroup of a free abelian group is a free abelian group, then $Z_p(K)$ and $B_p(K)$ are both free abelian groups. The quotient group

$$H_p(K) = Z_p(K)/B_p(K)$$

may not be free, but its possibilities are given by the decomposition theorem for finitely generated abelian groups (Appendix 3):

$$H_p(K) = G \oplus T_1 \oplus \cdots \oplus T_m$$

where G is a free abelian group and each T_i is a finite cyclic group. The direct sum $T_1 \oplus \cdots \oplus T_m$ is called the *torsion subgroup* of $H_p(K)$. As in the example with the projective plane, the torsion subgroup describes the "twisting" in the polyhedron $|K|$. Additional examples of twisting will be found in the exercises at the end of the chapter.

The existence of torsion subgroups explains why the integers modulo 2 are not generally used for the coefficient set in homology theory. The finite cyclic groups $T_1, \ldots, T_m$ which compose the torsion subgroup are quotient groups of $\mathbb{Z}$. If we used the group $\mathbb{Z}_2$ of integers modulo 2 rather than $\mathbb{Z}$, there would be no way to recognize torsion since $\mathbb{Z}_2$ admits no proper subgroups. Note also that orientation is meaningless in the modulo 2 case. For problems in which orientation and the torsion subgroup are not important, the integers modulo 2 can be an effective choice for the coefficient group. In this regard, see the chapter on modulo 2 homology theory, including the Jordan Curve Theorem, in [15].

The next theorem shows that the homology groups of a complex are independent of the choice of orientation for its simplexes.

Theorem 2.3. *Let K be a geometric complex with two orientations, and let K_1, K_2 denote the resulting oriented geometric complexes. Then the homology groups $H_p(K_1)$ and $H_p(K_2)$ are isomorphic for each dimension p.*

Proof. For a p-simplex σ^p of K, let ${}^i\sigma^p$ denote the positive orientation of σ^p in the complex K_i, $i = 1, 2$. Then there is a function α defined on the simplexes of K such that $\alpha(\sigma^p)$ is ± 1 and

$$ {}^1\sigma^p = \alpha(\sigma^p)\,{}^2\sigma^p. $$

Define a sequence $\varphi = \{\varphi_p\}$ of homomorphisms

$$ \varphi_p\colon C_p(K_1) \to C_p(K_2) $$

by

$$ \varphi_p\left(\sum g_i \cdot {}^1\sigma_i^p\right) = \sum \alpha(\sigma_i^p) g_i \cdot {}^2\sigma_i^p $$

where $\sum g_i \cdot {}^1\sigma_i^p$ represents a p-chain on K_1.

For an elementary p-chain $g \cdot {}^1\sigma^p$ on K_1 with $p \geq 1$,

$$ \begin{aligned} \varphi_{p-1}\partial(g \cdot {}^1\sigma^p) &= \varphi_{p-1}\left(\sum_{\sigma^{p-1}\in K} g[{}^1\sigma^p, {}^1\sigma^{p-1}] \cdot {}^1\sigma^{p-1}\right) \\ &= \sum_{\sigma^{p-1}\in K} \alpha(\sigma^{p-1}) g[{}^1\sigma^p, {}^1\sigma^{p-1}] \cdot {}^2\sigma^{p-1} \\ &= \sum_{\sigma^{p-1}\in K} \alpha(\sigma^{p-1}) g \alpha(\sigma^{p-1})\alpha(\sigma^p)[{}^2\sigma^p, {}^2\sigma^{p-1}] \cdot {}^2\sigma^{p-1} \\ &= \alpha(\sigma^p) g \sum_{\sigma^{p-1}\in K} [{}^2\sigma^p, {}^2\sigma^{p-1}] \cdot {}^2\sigma^{p-1} = \partial(\alpha(\sigma^p) g \cdot {}^2\sigma^p) \\ &= \partial\varphi_p(g \cdot {}^1\sigma^p). \end{aligned} $$

Thus the relation $\varphi_{p-1}\partial = \partial\varphi_p$ holds in the diagram

$$\begin{array}{ccc} C_p(K_1) & \xrightarrow{\varphi_p} & C_p(K_2) \\ \downarrow \partial & & \downarrow \partial \\ C_{p-1}(K_1) & \xrightarrow{\varphi_{p-1}} & C_{p-1}(K_2). \end{array}$$

(As we shall see later, this is a very important relation.) If $z_p \in Z_p(K_1)$, then

$$\partial\varphi_p(z_p) = \varphi_{p-1}\partial(z_p) = \varphi_{p-1}(0) = 0,$$

so $\varphi_p(z_p) \in Z_p(K_2)$. Hence $\varphi_p(Z_p(K_1))$ is a subset of $Z_p(K_2)$.

If $\partial(c_{p+1}) \in B_p(K_1)$, then

$$\varphi_p\partial(c_{p+1}) = \partial\varphi_{p+1}(c_{p+1}),$$

so $\varphi_p\partial(c_{p+1})$ is in $B_p(K_2)$. Thus φ_p maps $B_p(K_1)$ into $B_p(K_2)$ and induces a homomorphism φ_p^* from the quotient group $H_p(K_1) = Z_p(K_1)/B_p(K_1)$ to $H_p(K_2) = Z_p(K_2)/B_p(K_2)$ defined by

$$\varphi_p^*([z_p]) = [\varphi_p(z_p)]$$

for each homology class $[z_p]$ in $H_p(K_1)$.

Reversing the roles of K_1 and K_2 yields a sequence $\psi = \{\psi_p\}$ of homomorphisms:

$$\psi_p\colon C_p(K_2) \to C_p(K_1)$$

such that φ_p and ψ_p are inverses of each other for each p. This implies that ψ_p^* is the inverse of φ_p^* and hence that

$$\varphi_p^*\colon H_p(K_1) \to H_p(K_2)$$

is an isomorphism for each dimension p. □

As remarked earlier, the structure of the zero dimensional homology group $H_0(K)$ indicates whether or not the polyhedron $|K|$ is connected. Actually the situation is quite simple; there is no torsion in dimension zero, and the rank of the free abelian group $H_0(K)$ is the number of components of the polyhedron $|K|$. Proving this is our next goal.

Definition. Let K be a complex. Two simplexes s_1 and s_2 are *connected* if either of the following conditions is satisfied:

(a) $s_1 \cap s_2 \neq \varnothing$;
(b) there is a sequence $\sigma_1, \ldots, \sigma_p$ of 1-simplexes of K such that $s_1 \cap \sigma_1$ is a vertex of s_1, $s_2 \cap \sigma_p$ is a vertex of s_2, and, for $1 \leq i < p$, $\sigma_i \cap \sigma_{i+1}$ is a common vertex of σ_i and σ_{i+1}.

This concept of connectedness is an equivalence relation whose equivalence classes are called the *combinatorial components* of K. The complex K is said to be *connected* if it has only one combinatorial component.

It is left as an exercise for the reader to show that the components of $|K|$ and the geometric carriers of the combinatorial components of K are identical.

Theorem 2.4. *Let K be a complex with r combinatorial components. Then $H_0(K)$ is isomorphic to the direct sum of r copies of the group $\mathbb{Z}$ of integers.*

PROOF. Let K' be a combinatorial component of K and $\langle a' \rangle$ a 0-simplex in K'. Given any 0-simplex $\langle b \rangle$ in K', there is a sequence of 1-simplexes

$$\langle ba_0 \rangle, \langle a_0 a_1 \rangle, \langle a_1 a_2 \rangle, \ldots, \langle a_p a' \rangle$$

from b to a' such that each two successive 1-simplexes have a common vertex. If g is an integer, we define a 1-chain c_1 on the sequence of 1-simplexes by assigning either g or $-g$ to each simplex (depending on orientation) so that $\partial(c_1)$ is $g \cdot \langle b \rangle - g \cdot \langle a' \rangle$ or $g \cdot \langle b \rangle + g \cdot \langle a' \rangle$. Hence any elementary 0-chain $g \cdot \langle b \rangle$ is homologous to one of the 0-chains $g \cdot \langle a' \rangle$ or $-g \cdot \langle a' \rangle$. It follows that any 0-chain on K' is homologous to an elementary 0-chain $h \cdot \langle a' \rangle$ where h is some integer.

Applying this result to each combinatorial component $K_1, \ldots, K_r$ of K, there is a vertex a^i of K_i such that any 0-cycle on K_i is homologous to a 0-chain of the form $h_i \cdot \langle a^i \rangle$ where h_i is an integer. Then, given any 0-cycle c_0 on K, there are integers $h_1, \ldots, h_r$ such that

$$c_0 \sim \sum_{i=1}^{r} h_i \cdot \langle a^i \rangle.$$

Suppose that two such 0-chains $\sum h_i \cdot \langle a^i \rangle$ and $\sum g_i \cdot \langle a^i \rangle$ represent the same homology class. Then

$$\sum (g_i - h_i) \langle a^i \rangle = \partial(c_1) \tag{6}$$

for some 1-chain c_1. Since a^i and a^j belong to different combinatorial components when $i \neq j$, then Equation (6) is impossible unless $g_i = h_i$ for each i. Hence each homology class $[c_0]$ in $H_0(K)$ has a unique representative of the form $\sum h_i \cdot \langle a^i \rangle$. The function

$$\sum h_i \cdot \langle a^i \rangle g \rightarrow (h_1, \ldots, h_r)$$

is the required isomorphism between $H_0(K)$ and the direct sum of r copies of $\mathbb{Z}$. □

Corollary. *If a polyhedron $|K|$ has r components and triangulation K, then $H_0(K)$ is isomorphic to the direct sum of r copies of $\mathbb{Z}$.*

2.4 The Euler–Poincaré Theorem

If $|K|$ is a rectilinear polyhedron homeomorphic to the 2-sphere S^2 with V vertices, E edges, and F two dimensional faces, then

$$V - E + F = 2.$$

This result was discovered in 1752 by Leonhard Euler (1707–1783). Poincaré's first real application of homology theory was a generalization of Euler's formula to general polyhedra. That celebrated result, the Euler–Poincaré Theorem, is proved in this section.

Definition. Let K be an oriented complex. A family $\{z_p^1, \ldots, z_p^r\}$ of p-cycles is *linearly independent with respect to homology*, or *linearly independent* mod $B_p(K)$, means that there do not exist integers $g_1, \ldots, g_r$ not all zero such that the chain $\sum g_i z_p^i$ is homologous to 0. The largest integer r for which there exist r p-cycles linearly independent with respect to homology is denoted by $R_p(K)$ and called the pth *Betti number* of the complex K.

In the theorem that follows, we assume that the coefficient group has been chosen to be the rational numbers and not the integers. (This is one of two instances in which this change is made.) The reader should convince himself that linear independence with integral coefficients is equivalent to linear independence with rational coefficients and that this change does not alter the values of the Betti numbers.

Theorem 2.5. (The Euler–Poincaré Theorem). *Let K be an oriented geometric complex of dimension n, and for $p = 0, 1, \ldots, n$ let α_p denote the number of p-simplexes of K. Then*

$$\sum_{p=0}^{n} (-1)^p \alpha_p = \sum_{p=0}^{n} (-1)^p R_p(K)$$

where $R_p(K)$ denotes the pth Betti number of K.

Proof. Since K is the only complex under consideration, the notation will be simplified by omitting reference to it in the group notations. Note that C_p, Z_p, and B_p are vector spaces over the field of rational numbers.

Let $\{d_p^i\}$ be a maximal set of p-chains such that no proper linear combination of the d_p^i is a cycle, and let D_p be the linear subspace of C_p spanned by $\{d_p^i\}$. Then $D_p \cap Z_p = \{0\}$ so that, as a vector space, C_p is the direct sum of Z_p and D_p. Hence

$$\alpha_p = \dim C_p = \dim D_p + \dim Z_p,$$
$$\dim Z_p = \alpha_p - \dim D_p, \qquad 1 \leq p \leq n,$$

where the abbreviation "dim" denotes vector space dimension.

For $p = 0, \ldots, n - 1$, let $b_p^i = \partial(d_{p+1}^i)$. The set $\{b_p^i\}$ forms a basis for B_p. Let $\{z_p^i\}$, $i = 1, \ldots, R_p$, be a maximal set of p-cycles linearly independent mod B_p. These cycles span a subspace G_p of Z_p, and

$$Z_p = G_p \oplus B_p, \qquad 0 \leq p \leq n - 1.$$

Thus

$$\dim Z_p = \dim G_p + \dim B_p = R_p + \dim B_p$$

since $R_p = \dim G_p$. Then

$$R_p = \dim Z_p - \dim B_p = \alpha_p - \dim D_p - \dim B_p, \qquad 1 \leq p \leq n - 1.$$

Observe that B_p is spanned by the boundaries of elementary chains

$$\partial(1\cdot\sigma_i^{p+1}) = \sum \eta_{ij}(p)\cdot\sigma_j^p$$

where $(\eta_{ij}(p)) = \eta(p)$ is the pth incidence matrix. Thus $\dim B_p = \operatorname{rank}\eta(p)$. Since the number of d_{p+1}^i is the same as the number of b_p^i, then

$$\dim D_{p+1} = \dim B_p = \operatorname{rank}\eta(p), \qquad 0 \le p \le n-1.$$

Then

$$\begin{aligned} R_p &= \alpha_p - \dim D_p - \dim B_p \\ &= \alpha_p - \operatorname{rank}\eta(p-1) - \operatorname{rank}\eta(p), \qquad 1 \le p \le n-1. \end{aligned}$$

Note also that

$$\begin{aligned} R_0 &= \dim Z_0 - \dim B_0 = \alpha_0 - \operatorname{rank}\eta(0) \\ R_n &= \dim Z_n = \alpha_n - \dim D_n = \alpha_n - \operatorname{rank}\eta(n-1). \end{aligned}$$

In the alternating sum $\sum_{p=0}^n (-1)^p R_p$, all the terms rank $\eta(p)$ cancel, and we have

$$\sum_{p=0}^{n} (-1)^p R_p = \sum_{p=0}^{n} (-1)^p \alpha_p. \qquad \square$$

Definition. If K is a complex of dimension n, the number

$$\chi(K) = \sum_{p=0}^{n} (-1)^p R_p$$

is called the *Euler characteristic* of K.

Chains, cycles, boundaries, the homology relation, and Betti numbers were defined by Poincaré in his paper *Analysis Situs* [49] in 1895. As mentioned earlier, he did not define the homology groups. The proof of the Euler–Poincaré Theorem given in the text is essentially Poincaré's original one. Complexes (in slightly different form) and incidence numbers were defined in *Complément à l'Analysis Situs* [50] in 1899.

The Betti numbers were named for Enrico Betti (1823–1892) and generalize the connectivity numbers that he used in studying curves and surfaces. Poincaré assumed, but did not prove, that the Betti numbers are topological invariants. In other words, he assumed that if the geometric carriers $|K|$ and $|L|$ are homeomorphic, then $R_p(K) = R_p(L)$ in each dimension p. The first rigorous proof of this fact was given by J. W. Alexander (1888–1971) in 1915. Topological invariance of the homology groups was proved by Oswald Veblen in 1922. One can thus speak of $H_p(|K|)$, $R_p(|K|)$, and $\chi(|K|)$ since these homology characters are independent of the triangulation of the polyhedron $|K|$. It is important to know that the homology characters are topologically invariant. The proofs are lengthy, however, and are omitted. Anyone interested in following this topic further should consult references [2] and [17].

It is left as an exercise to show that the pth Betti number $R_p(K)$ of a complex K is the rank of the free part of the pth homology group $H_p(K)$. The pth Betti number indicates the number of "p-dimensional holes" in the polyhedron $|K|$.

Definition. A *rectilinear polyhedron* in Euclidean 3-space $\mathbb{R}^3$ is a solid bounded by properly joined convex polygons. The bounding polygons are called *faces*, the intersections of the faces are called *edges*, and the intersections of the edges are called *vertices*. A *simple polyhedron* is a rectilinear polyhedron whose boundary is homeomorphic to the 2-sphere S^2. A *regular polyhedron* is a rectilinear polyhedron whose faces are regular plane polygons and whose polyhedral angles are congruent.

In Exercise 6 at the end of the chapter, the reader will find that the Betti numbers of the 2-sphere S^2 are

$$R_0(S^2) = 1, \qquad R_1(S^2) = 0, \qquad R_2(S^2) = 1.$$

Then S^2 has Euler characteristic

$$\chi(S^2) = \sum_{p=0}^{2} (-1)^p R_p(S^2) = 1 - 0 + 1 = 2.$$

Applying the Euler–Poincaré Theorem to S^2 produces the following corollary:

Theorem 2.6 (Euler's Theorem). *If S is a simple polyhedron with V vertices, E edges, and F faces, then $V - E + F = 2$.*

PROOF. Things are complicated slightly here by the fact that the faces of S need not be triangular. This situation is corrected as follows: Consider a face τ of S having n_0 vertices and n_1 edges. Computing *vertices* $-$ *edges* $+$ *faces* gives $n_0 - n_1 + 1$ for the single face τ. Choose a new vertex v in the interior of τ, and join the new vertex to each of the original vertices by a line segment as illustrated in Figure 2.3. In the triangulation of τ, one new vertex and n_0

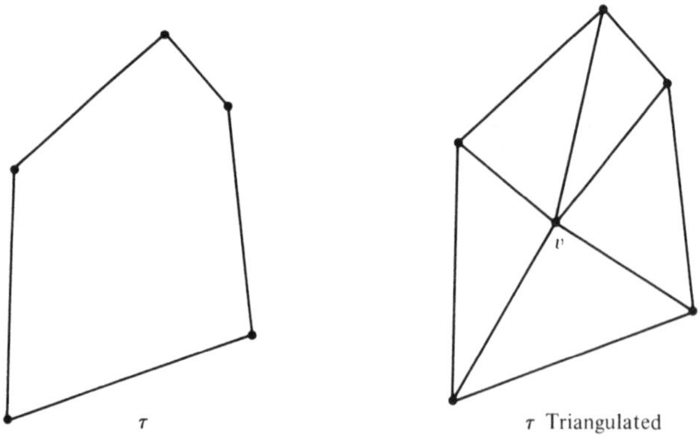

Figure 2.3

new edges are added. In addition, the one face τ is replaced by n_0 new faces. Then

$$\text{vertices} - \text{edges} + \text{faces} = (n_0 + 1) - (n_1 + n_0) + n_0 = n_0 - n_1 + 1$$

so that the sum $V - E + F$ is not changed in the triangulation process. Let α_i, $i = 0, 1, 2$, denote the number of i-simplexes in the triangulation of S obtained in this way. Then

$$V - E + F = \alpha_0 - \alpha_1 + \alpha_2$$

by the above argument. The Euler–Poincaré Theorem shows that

$$\alpha_0 - \alpha_1 + \alpha_2 = R_0(S^2) - R_1(S^2) + R_2(S^2) = 2.$$

Hence

$$V - E + F = 2$$

for any simple polyhedron. □

Theorem 2.7. *There are only five regular, simple polyhedra.*

Proof. Suppose S is such a polyhedron with V vertices, E edges, and F faces. Let m denote the number of edges meeting at each vertex and n the number of edges of each face. Note that $n \geq 3$. Then

$$mV = 2E = nF,$$
$$V - E + F = 2$$

so that

$$\frac{nF}{m} - \frac{nF}{2} + F = 2.$$

Hence

$$F(2n - mn + 2m) = 4m,$$

and it must be true that

$$2n - mn + 2m > 0.$$

Since $n \geq 3$, this gives

$$2m > n(m - 2) \geq 3(m - 2) = 3m - 6,$$

so $m < 6$. Thus m can only be 1, 2, 3, 4, or 5.

The relations

$$F(2n - mn + 2m) = 4m, \qquad n \geq 3, m < 6$$

produce the following possible values for (m, n, F): (a) $(3, 3, 4)$, (b) $(3, 4, 6)$, (c) $(4, 3, 8)$, (d) $(3, 5, 12)$, and (e) $(5, 3, 20)$.

For example, $m = 4$ gives

$$F(8 - 2n) = 16,$$

allowing the possibility $F = 8$, $n = 3$. (The reader should solve the remaining cases.) The five possibilities for (m, n, F) are realized in the tetrahedron, cube, octahedron, dodecahedron, and icosahedron shown in Figure 2.4. □

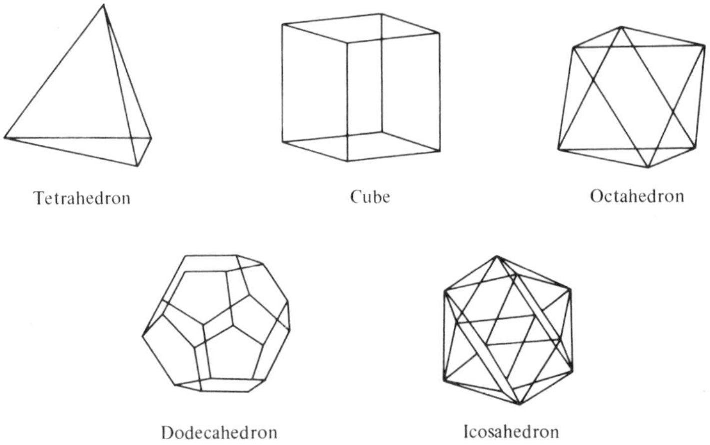

Figure 2.4

2.5 Pseudomanifolds and the Homology Groups of S^n

Algebraic topology developed from problems in mathematical analysis and geometry in Euclidean spaces, particularly Poincaré's work in the classification of algebraic surfaces. The spaces of primary interest, called "manifolds", can be traced to the work of G. F. B. Riemann (1826–1866) on differentials and multivalued functions. A manifold is a generalization of an ordinary surface like a sphere or a torus; its primary characteristic is its "local" Euclidean structure. Here is the definition:

Definition. An *n-dimensional manifold*, or *n-manifold*, is a compact, connected Hausdorff space each of whose points has a neighborhood homeomorphic to an open ball in Euclidean n-space $\mathbb{R}^n$.

It should be noted that not all texts require that manifolds be compact and connected. Sometimes these conditions are omitted, and other properties, paracompactness and second countability, for example, are added. For many of the applications in this text, however, compactness and connectedness are required, and it will simplify matters to include them in the definition.

Definition. An *n-pseudomanifold* is a complex K with the following properties:

(a) Each simplex of K is a face of some n-simplex of K.
(b) Each $(n - 1)$-simplex is a face of exactly two n-simplexes of K.
(c) Given a pair σ_1^n and σ_2^n of n-simplexes of K, there is a sequence of n-simplexes beginning with σ_1^n and ending with σ_2^n such that any two successive terms of the sequence have a common $(n - 1)$-face.

Example 2.4. (a) The complex K consisting of all proper faces of a 3-simplex $\langle a_0a_1a_2a_3\rangle$ (Figure 2.5) is a 2-pseudomanifold and is a triangulation of the 2-sphere S^2.

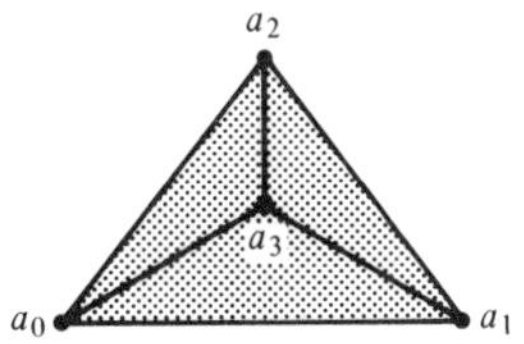

Figure 2.5

(b) The triangulation of the projective plane in Figure 2.2 is a 2-pseudomanifold.

(c) The triangulation of the torus in Figure 1.11 is a 2-pseudomanifold.

(d) The *Klein Bottle* is constructed from a cylinder by identifying opposite ends with the orientations of the circles reversed. A triangulation of the Klein Bottle as a 2-pseudomanifold is shown in Figure 2.6.

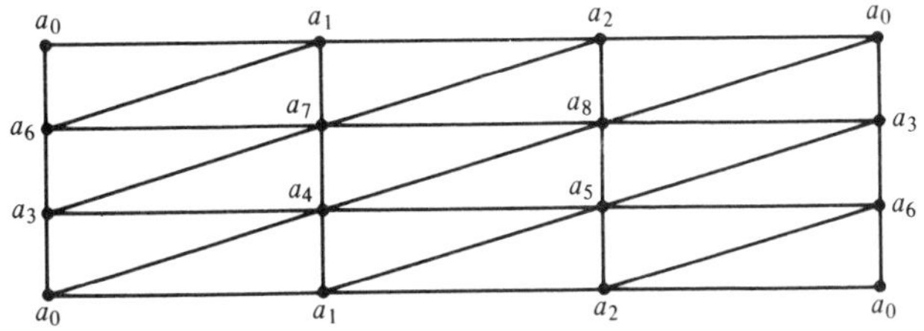

Figure 2.6 *Triangulation of the Klein Bottle*

The Klein Bottle cannot be embedded in Euclidean 3-space without self-intersection. Allowing self-intersection, it appears in the figure below.

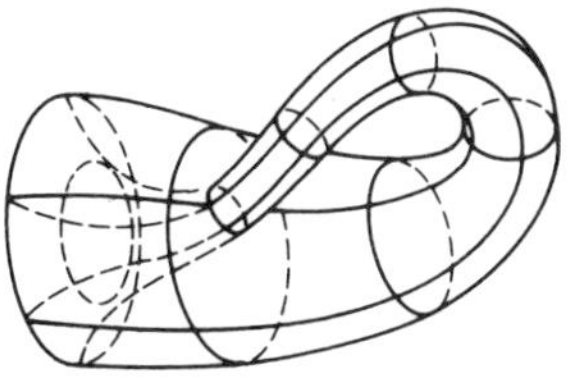

Figure 2.7

Each space of Example 2.4 is a 2-manifold. The n-sphere S^n, $n \geq 1$, is an n-manifold. Incidentally, this indicates why the unit sphere in $\mathbb{R}^{n+1}$ is called the "n-sphere" and not the "$(n + 1)$-sphere". The integer n refers to the local dimension as a manifold and not to the dimension of the containing Euclidean space. Note that each point of a circle has a neighborhood homeomorphic to an open interval in $\mathbb{R}$; each point of S^2 has a neighborhood homeomorphic to an open disk in $\mathbb{R}^2$; and so on.

The relation between manifold (a type of topological space) and pseudomanifold (a type of geometric complex) is simple to state: If X is a triangulable n-manifold, then each triangulation K of X is an n-pseudomanifold. The homology groups of the pseudomanifold K reflect the connectivity, the "holes" and "twisting", of the associated manifold X. The computation of homology groups of pseudomanifolds is thus a worthwhile project. As we shall see in this section, these groups are often amenable to computation.

If X is a space each of whose triangulations is a pseudomanifold, it is sometimes said that "X is a pseudomanifold." Since a space and a triangulation of the space are different, this is an abuse of language. It is permissible only in situations in which the distinction between space and complex is not important, as in the computation of homology groups.

We shall restrict ourselves in this section to theorems and examples related to the homology groups of pseudomanifolds. Those interested in the fact that each triangulation of a triangulable n-manifold is an n-pseudomanifold can find the proof in many texts, for example [2].

Theorem 2.8. *Let K be a 2-pseudomanifold with α_0 vertices, α_1 1-simplexes, and α_2 2-simplexes. Then*

(a) $3\alpha_2 = 2\alpha_1$,

(b) $\alpha_1 = 3(\alpha_0 - \chi(K))$,

(c) $\alpha_0 \geq \frac{1}{2}(7 + \sqrt{49 - 24\chi(K)})$.

Proof. Since each 1-simplex is a face of exactly two 2-simplexes, it follows that $3\alpha_2 = 2\alpha_1$ and hence that $\alpha_2 = \frac{2}{3}\alpha_1$.

The Euler–Poincaré Theorem guarantees that

$$\alpha_0 - \alpha_1 + \alpha_2 = \chi(K).$$

Then

$$\alpha_0 - \alpha_1 + \tfrac{2}{3}\alpha_1 = \chi(K),$$

and hence

$$\alpha_1 = 3(\alpha_0 - \chi(K)).$$

To prove (c), note that $\alpha_0 \geq 4$ and that

$$\alpha_1 \leq C_2^{\alpha_0} = \tfrac{1}{2}\alpha_0(\alpha_0 - 1)$$

where $C_2^{\alpha_0}$ denotes the number of combinations of α_0 vertices taken two at a time. By elementary algebra,

$$\begin{aligned}
6\alpha_2 &= 4\alpha_1 \\
2\alpha_1 &= 6\alpha_1 - 6\alpha_2 \\
\alpha_0(\alpha_0 - 1) &\geq 6\alpha_1 - 6\alpha_2 \\
\alpha_0^2 - \alpha_0 - 6\alpha_0 &\geq 6\alpha_1 - 6\alpha_2 - 6\alpha_0 = -6\chi(K) \\
\alpha_0^2 - 7\alpha_0 &\geq -6\chi(K) \\
4\alpha_0^2 - 28\alpha_0 + 49 &\geq 49 - 24\chi(K) \\
(2\alpha_0 - 7)^2 &\geq 49 - 24\chi(K) \\
\alpha_0 &\geq \tfrac{1}{2}(7 + \sqrt{49 - 24\chi(K)}).
\end{aligned}$$

□

Theorem 2.8 is useful in determining the 2-pseudomanifold triangulation of a polyhedron having the minimum number of simplexes in each dimension. Computing homology groups is at best a tedious procedure; it is simplified by using a minimal triangulation (a triangulation with the smallest number of simplexes).

Example 2.5. Consider, for example, the 2-sphere S^2. Since $\chi(S^2) = 2$, then

$$\begin{aligned}\alpha_0 &\geq \tfrac{1}{2}(7 + \sqrt{49 - 24\chi(K)}) = 4,\\ \alpha_1 &= 3(\alpha_0 - \chi(K)) \geq 3(4 - 2) = 6,\\ \alpha_2 &= \tfrac{2}{3}\alpha_1 \geq \tfrac{2}{3}\cdot 6 = 4.\end{aligned}$$

Hence any triangulation of S^2 must have at least four vertices, at least six 1-simplexes, and at least four 2-simplexes. This minimal triangulation is achieved by the boundary complex of a tetrahedron (proper faces of a 3-simplex) in Figure 2.5.

Example 2.6. Consider the projective plane P, a 2-manifold. As shown earlier, $H_2(P) = \{0\}$ and $H_1(P) \cong \mathbb{Z}_2$. Since P is connected, Theorem 2.4 shows that $H_0(P) \cong \mathbb{Z}$. Then

$$R_2(P) = R_1(P) = 0, \qquad R_0(P) = 1, \qquad \chi(P) = 1.$$

This gives

$$\begin{aligned}\alpha_0 &\geq \tfrac{1}{2}(7 + \sqrt{49 - 24\chi(P)}) = 6,\\ \alpha_1 &\geq 3(6 - 1) = 15,\\ \alpha_2 &\geq \tfrac{2}{3}\cdot 15 = 10,\end{aligned}$$

so that any triangulation of P must have at least six vertices, fifteen 1-simplexes, and ten 2-simplexes. The triangulation of P given in Figure 2.2 is thus minimal.

Definition. Let K be an n-pseudomanifold. For each $(n - 1)$-simplex σ^{n-1} of K, let σ_1^n and σ_2^n denote the two n-simplexes of which σ^{n-1} is a face. An orientation for K having the property

$$[\sigma_1^n, \sigma^{n-1}] = -[\sigma_2^n, \sigma^{n-1}]$$

for each $(n - 1)$-simplex σ^{n-1} of K is a *coherent orientation*. An n-pseudomanifold is *orientable* if it can be assigned a coherent orientation. Otherwise it is *nonorientable*.

The proof is lengthy, but it can be shown that orientability is a topological property of the underlying polyhedron $|K|$ and is not dependent on the particular triangulation K. We shall assume this without proof. It is left as an exercise for the reader to show that the projective plane and Klein Bottle are nonorientable while the 2-sphere and torus are orientable.

Example 2.7. Let K denote the n-skeleton of the closure of an $(n + 1)$-simplex σ^{n+1} in $\mathbb{R}^{n+1}$, $n \geq 1$. Then K is an n-pseudomanifold and is a triangulation of the n-sphere S^n. (Recall Exercise 12 in Chapter 1.)

The following notation will be helpful in determining a coherent orientation and is used only in this example. For an integer j with $0 \leq j \leq n + 1$, let

$$\sigma_j = \langle a_0 \ldots \hat{a}_j \ldots a_{n+1} \rangle$$

where the symbol $\hat{a}_j$ indicates that the vertex a_j is deleted. The positively oriented simplex $+\sigma_j$ has the given ordering when j is even and the opposite ordering (an odd permutation of the given ordering) when j is odd. The $(n - 1)$-simplex

$$+\sigma_{ij} = +\langle a_0 \ldots \hat{a}_i \ldots \hat{a}_j \ldots a_{n+1} \rangle$$

is then a face of the two n-simplexes σ_i and σ_j.

It is left as an exercise for the reader to show that this orientation for the n-simplexes and $(n - 1)$-simplexes gives

$$[\sigma_i, \sigma_{ij}] = -[\sigma_j, \sigma_{ij}]$$

in each case. It follows that any n-chain of the form $\sum_{\sigma_i \in K} g \cdot \sigma_i$, g an integer, is an n-cycle. Furthermore, if

$$z = \sum_{\sigma_i \in K} g_i \cdot \sigma_i$$

is an n-cycle, then

$$0 = \partial(z) = \sum_{\sigma_{ij} \in K} h_{ij} \cdot \sigma_{ij}$$

where h_{ij} is either $g_i - g_j$ or $g_j - g_i$. Hence z is an n-cycle if and only if all the coefficients g_i have a common value g. Thus $Z_n(S^n) \cong \mathbb{Z}$. Since $B_n(S^n) = \{0\}$, then $H_n(S^n) \cong \mathbb{Z}$.

A complete description of the homology groups of S^n is given by the following theorem:

Theorem 2.9. *The homology groups of the n-sphere, $n \geq 1$, are*

$$H_p(S^n) \cong \begin{cases} \mathbb{Z} & \text{if } p = 0 \text{ or } p = n \\ \{0\} & \text{if } 0 < p < n. \end{cases}$$

PROOF. Since S^n is connected, Theorem 2.4 implies that $H_0(S^n) \cong \mathbb{Z}$. The above example shows that $H_n(S^n) \cong Z$. The following notation will be used in handling the case $0 < p < n$: If $+\sigma^p = \langle a_0 \ldots a_p \rangle$ and v is a vertex for which the set $\{v, a_0, \ldots, a_p\}$ is geometrically independent, then the symbol $v\sigma^p$ denotes the positively oriented $(p + 1)$-simplex $+\langle va_0 \ldots a_p \rangle$. If $c = \sum g_i \cdot \sigma_i^p$ is a p-chain, then vc denotes the $(p + 1)$-chain

$$vc = \sum g_i \cdot v\sigma_i^p.$$

Note that

$$\partial(1 \cdot v\sigma^p) = 1 \cdot \sigma^p - v\partial(1 \cdot \sigma^p).$$

Now consider a particular vertex v in the triangulation of S^n given in the preceding example. Since any p-simplex containing v can be expressed in the form $v\sigma^{p-1}$, then any p-cycle z can be written

$$z = \sum g_i \cdot \sigma_i^p + \sum h_j \cdot v\sigma_j^{p-1}$$

where simplexes in the second sum have v as a vertex and those in the first sum do not. Since z is a p-cycle, then

$$\begin{aligned} 0 = \partial(z) &= \partial\left(\sum g_i \cdot \sigma_i^p\right) + \partial\left(\sum h_j \cdot v\sigma_j^{p-1}\right) \\ &= \partial\left(\sum g_i \cdot \sigma_i^p\right) + \sum h_j \cdot \sigma_j^{p-1} - v\left(\partial \sum h_j \cdot \sigma_j^{p-1}\right) \end{aligned}$$

so that

$$\partial\left(\sum h_j \cdot \sigma_j^{p-1}\right) = 0, \qquad \partial\left(\sum g_i \cdot \sigma_i^p\right) = -\sum h_j \cdot \sigma_j^{p-1}.$$

This gives

$$\begin{aligned} \partial\left(\sum g_i v\sigma_i^p\right) &= \sum g_i \cdot \sigma_i^p - v\partial\left(\sum g_i \cdot \sigma_i^p\right) \\ &= \sum g_i \cdot \sigma_i^p + v \sum h_j \cdot \sigma_j^{p-1} = z. \end{aligned}$$

Thus every p-cycle on S^n is a boundary, so $H_p(S^n) = \{0\}$ for $0 < p < n$. □

The next theorem explains the meaning of orientability in terms of homology groups.

Theorem 2.10. *An n-pseudomanifold K is orientable if and only if the nth homology group $H_n(K)$ is not the trivial group.*

Proof. Assume first that K is orientable and assign it a coherent orientation. Then if the $(n-1)$-simplex σ^{n-1} is a face of σ_1^n and σ_2^n, we have

$$[\sigma_1^n, \sigma^{n-1}] = -[\sigma_2^n, \sigma^{n-1}].$$

This implies that any n-chain of the form

$$c = \sum_{\sigma^n \in K} g \cdot \sigma^n$$

(g a fixed integer) is an n-cycle. Thus $Z_n(K) \neq \{0\}$. Since $B_n(K) = \{0\}$, then $H_n(K) \neq \{0\}$.

To complete the proof it must be shown that K is orientable if $H_n(K) \neq \{0\}$. Suppose that

$$z = \sum_{\sigma_i^n \in K} g_i \cdot \sigma_i^n$$

is a nonzero n-cycle.

Since each pair of n-simplexes in K can be joined by a sequence of n-simplexes (as specified in the definition of n-pseudomanifold) and each $(n-1)$-simplex is a face of exactly two n-simplexes, it follows that any two coefficients in z can differ only in sign. That is to say, $g_i = \pm g_0$ if $\partial(z) = 0$. By reorienting σ_i^n if $g_i = -g_0$, we obtain an n-cycle

$$\sum_{\sigma_i^n \in K} g_0 \cdot \sigma_i^n = g_0\left(\sum_{\sigma_i^n \in K} 1 \cdot \sigma_i^n\right),$$

so it follows that $\sum 1 \cdot \sigma_i^n$ is an n-cycle. But this means that each $(n-1)$-simplex must have positive incidence number with one of the n-simplexes of which it is a face and negative incidence number with the other. In other words, K is orientable. □

Corollary. *An n-pseudomanifold L is nonorientable if and only if* $H_n(L) = \{0\}$.

The question of whether or not every n-manifold has a triangulation was raised by Poincaré. Here it was not required that manifolds be compact, and triangulations having an infinite number of simplexes were allowed. Under these conditions, Tibor Rado (1895–1965) proved in 1922 that every 2-manifold has a triangulation, and Edwin Moise (1918–) proved the corresponding result for 3-manifolds in 1952.

In 1969 R. C. Kirby (1938–) and L. C. Siebenmann (1939–), using a somewhat different definition of triangulability, showed the existence of manifolds in higher dimensions which are not triangulable in their sense of the term. This answered a related triangulation problem which had been of interest for many years. The results of Kirby and Siebenmann can be found in [44].

A 2-manifold is called a *closed surface*. The topological power of the homology groups is demonstrated by the following classification theorem for closed surfaces.

Theorem 2.11. *Two closed surfaces are homeomorphic if and only if they have the same Betti numbers in corresponding dimensions.*

The proof of Theorem 2.11 is omitted from this text because it would require a lengthy digression into the theory of closed surfaces and because, historically, the theorem preceded Poincaré's formalization of algebraic topology. It was a motivating force behind Poincaré's work, however, and served as a model of the type of theorem to which topology would aspire. More will be said on this point in Chapter 4.

Theorem 2.11 was essentially known by about 1890 through the work of various mathematicians, notably Camille Jordan (1858–1922) and A. F. Möbius (1790–1860). Jordan is best known for his work in algebra and for proposing the Jordan Curve Theorem. Möbius invented the polyhedron that bears his name (the Möbius strip) and in so doing initiated the study of orientability. He used the term "one-sided" to mean nonorientable and "two-sided" to mean orientable for surfaces. The modern terms "orientable" and "nonorientable" were introduced by J. W. Alexander to generalize Möbius' concepts to higher dimensions.

Those who wish to see a proof of Theorem 2.11 should consult the texts by Cairns [2] or Massey [16].

EXERCISES

1. Suppose that K_1 and K_2 are two triangulations of the same polyhedron. Are the chain groups $C_p(K_1)$ and $C_p(K_2)$ isomorphic? Explain.

2. Suppose that complexes K_1 and K_2 have the same simplexes but different orientations. How are the chain groups $C_p(K_1)$ and $C_p(K_2)$ related?

3. Prove Theorem 2.2.

4. Let z_p be a p-cycle on a complex K. Explain why the homology class $[z_p]$ and the coset $z_p + B_p(K)$ are identical.

5. Let K denote the complex consisting of all proper faces of a 2-simplex $\langle a_0 a_1 a_2 \rangle$ with orientation induced by the order $a_0 < a_1 < a_2$. Compute all homology groups of K.

6. Compute the homology groups and Betti numbers of the 2-sphere S^2.

7. Compute the homology groups of the cylinder C triangulated in the accompanying figure. (Assign any orientation you like.)

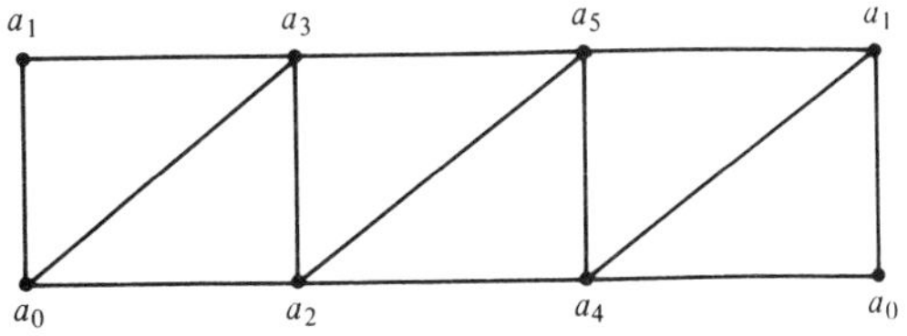

8. Compute the homology groups of the torus.

9. Compute the homology groups of the Klein Bottle.

10. Prove that linear independence with respect to homology for integral coefficients is equivalent to linear independence with respect to homology for rational coefficients. Explain in particular why the Betti numbers are not altered by the change to rational coefficients.

11. Derive the possibilities for (m, n, F) referred to in the proof of Theorem 2.7. How do you rule out the cases $m = 1$ and $m = 2$?

12. Fill in the details in the proof of Theorem 2.3. Explain in particular the relation between $[{}^1\sigma^p, {}^1\sigma^{p-1}]$ and $[{}^2\sigma^p, {}^2\sigma^{p-1}]$.

13. Prove that the geometric carriers of the combinatorial components of a complex K and the components of the polyhedron $|K|$ are identical.

14. Prove that the pth Betti number of a complex K is the rank of the free part of the pth homology group $H_p(K)$.

15. Find a minimal triangulation for the torus T. (Its homology groups are $H_0(T) \cong \mathbb{Z}$, $H_1(T) \cong \mathbb{Z} \oplus \mathbb{Z}$, and $H_2(T) \cong \mathbb{Z}$.)

16. Let K be a complex and K^r its r-skeleton. Show that $H_p(K)$ and $H_p(K^r)$ are isomorphic for $0 \leq p < r$. How are $H_r(K)$ and $H_r(K^r)$ related?

17. Why must an n-pseudomanifold have dimension n?

18. Show explicitly that the torus is orientable and that the projective plane and Klein Bottle are nonorientable.

19. Complete the proof in Example 2.7 that the n-sphere S^n is orientable.

20. In the proof of Theorem 2.9, show that

$$\partial(1 \cdot v\sigma^p) = 1 \cdot \sigma^p - v\,\partial(1 \cdot \sigma^p).$$

21. Let K denote the closure of an n-simplex. Prove that $H_p(K) = \{0\}$ for $0 < p \leq n$. Use this to show that $H_p(S^n) = \{0\}$ for $0 < p < n$.

22. Show that an orientable n-pseudomanifold has exactly two coherent orientations for its n-simplexes.

23. If K is an orientable n-pseudomanifold, prove that $H_n(K) \cong \mathbb{Z}$.

24. In the definition of n-pseudomanifold, replace (b) with (b′): Each $(n-1)$-simplex is a face of at least one and at most two n-simplexes. The resulting conditions (a), (b′), and (c) define the term *n-pseudomanifold with boundary.*

(i) Define orientability for n-pseudomanifolds with boundary in analogy with the definition of orientability for n-pseudomanifolds.

(ii) Show that the Möbius strip is a nonorientable 2-pseudomanifold with boundary.

25. If K is a 2-pseudomanifold, prove that $\chi(K) \leq 2$. How is this fact used in Theorem 2.8?

26. Show that the projective plane P is the quotient space of the 2-sphere obtained by identifying each pair x, $-x$ of diametrically opposite points.

27. References [9] and [2] may be helpful for (b) and (c).

(a) Define a 1-dimensional complex K in $\mathbb{R}^3$ for which $|K|$ is not homeomorphic to a subspace of $\mathbb{R}^2$.

(b) Prove that if K is a complex of dimension n, then $|K|$ can be rectilinearly imbedded in $\mathbb{R}^{2n+1}$.

(c) Prove that every triangulation of an n-manifold is an n-pseudomanifold.

Simplicial Approximation 3

3.1 Introduction

We turn now to the problem of comparing polyhedra by means of their associated homology groups. Comparisons between two topological spaces are usually made on the basis of a continuous map, ideally a homeomorphism, from one space to another. Groups are compared by means of homomorphisms and isomorphisms. We shall show in this chapter that a continuous map $f: |K| \to |L|$ induces for each non-negative integer p a homomorphism $f_p^*: H_p(K) \to H_p(L)$ on the associated homology groups. This will allow topological comparisons between the polyhedra $|K|$ and $|L|$ on the basis of algebraic similarities between their associated homology groups.

We have pointed out that if $|K|$ and $|L|$ are homeomorphic, then $H_p(K)$ and $H_p(L)$ are isomorphic in each dimension p. The reader should be warned that the converse is not true. Even if there is a continuous map $f: |K| \to |L|$ for which f_p^* is an isomorphism for each dimension p, it may not follow that $|K|$ and $|L|$ are homeomorphic. Thus we do not have the best possible situation in which a topological comparison is reduced to a purely algebraic one. However, as we shall see in this and later chapters, the method of comparing topological spaces through their homology groups is a very powerful tool.

Suppose then that there is a continuous map $f: |K| \to |L|$ from one polyhedron to another. How are the associated homomorphisms defined? The situation would be simple if f took simplexes of K to simplexes of L, i.e., if f were a "simplicial map." We could then induce homomorphisms from $C_p(K)$ to $C_p(L)$ and use these to define the required homomorphisms on the homology groups. If f does not take simplexes of K to simplexes of L, we replace f by a map which does as follows: Subdivide K into smaller simplexes so that f "almost" maps each simplex of K into a simplex of L. We can then define explicitly a simplicial map which has the essential characteristics of f and use this new map to induce homomorphisms on the homology groups. The process of subdividing K is called "barycentric subdivision," and the associated simplicial map is called a "simplicial approximation." This intuitive description will be made more precise as we proceed. The existence of simplicial approximations to any continuous map $f: |K| \to |L|$ is the central result of this chapter.

3.2 Simplicial Approximation

Definition. Let K and L be complexes and $\{\varphi_p\}_0^\infty$ a sequence of homomorphisms $\varphi_p: C_p(K) \to C_p(L)$ such that

$$\partial\varphi_p = \varphi_{p-1}\partial, \qquad p \geq 1.$$

Then $\{\varphi_p\}_0^\infty$ is called a *chain mapping* from K into L.

In the preceding definition, the sequence $\{\varphi_p\}_0^\infty$ is written as an infinite sequence simply to avoid mention of the dimensions of K and L. When p exceeds dim K and dim L, then $C_p(K)$ and $C_p(L)$ are zero groups and φ_p must be the trivial homomorphism which takes 0 to 0.

Theorem 3.1. *A chain mapping $\{\varphi_p\}_0^\infty$ from a complex K into a complex L induces homomorphisms*

$$\varphi_p^*: H_p(K) \to H_p(L)$$

in each dimension p.

PROOF. If $b_p = \partial(c_{p+1})$ in $B_p(K)$, then

$$\varphi_p(b_p) = \varphi_p\partial(c_{p+1}) = \partial\varphi_{p+1}(c_{p+1}),$$

so $\varphi_p(b_p)$ is the boundary of the $(p + 1)$-chain $\varphi_{p+1}(c_{p+1})$. Thus φ_p maps $B_p(K)$ into $B_p(L)$.

We shall now show that φ_p maps $Z_p(K)$ into $Z_p(L)$. This is true for $p = 0$ since $Z_0(K) = C_0(K)$ and $Z_0(L) = C_0(L)$. For $p \geq 1$, suppose that $z_p \in Z_p(K)$. Note that

$$\partial\varphi_p(z_p) = \varphi_{p-1}\partial(z_p) = \varphi_{p-1}(0) = 0,$$

so $\varphi_p(z_p)$ is a p-cycle on L.

Since

$$H_p(K) = Z_p(K)/B_p(K), \qquad H_p(L) = Z_p(L)/B_p(L),$$

then the induced homomorphism $\varphi_p^*: H_p(K) \to H_p(L)$ can be defined in the standard way:

$$\varphi_p^*(z_p + B_p(K)) = \varphi_p(z_p) + B_p(L)$$

or, equivalently,

$$\varphi_p^*([z_p]) = [\varphi_p(z_p)]. \qquad \square$$

Definition. A *simplicial mapping* from a complex K into a complex L is a function φ from the vertices of K into those of L such that if $\sigma^p = \langle v_0 \ldots v_p \rangle$ is a simplex of K, then the vertices $\varphi(v_i)$, $0 \leq i \leq p$ (not necessarily distinct) are the vertices of a simplex of L. If the vertices $\varphi(v_i)$ are all distinct, then the p-simplex $\langle \varphi(v_0) \ldots \varphi(v_p) \rangle = \varphi(\sigma^p)$ is called the *image* of σ^p. If $\varphi(v_i) = \varphi(v_j)$ for some $i \neq j$, then φ is said to *collapse* σ^p.

Definition. Let φ be a simplicial mapping from K into L and p a non-negative integer. If $g \cdot \sigma^p$ is an elementary p-chain on K, define

$$\varphi_p(g \cdot \sigma^p) = \begin{cases} 0 & \text{if } \varphi \text{ collapses } \sigma^p \\ g \cdot \varphi(\sigma^p) & \text{if } \varphi \text{ does not collapse } \sigma^p. \end{cases}$$

The function φ_p is extended by linearity to a homomorphism $\varphi_p \colon C_p(K) \to C_p(L)$. That is to say, if $\sum g_i \cdot \sigma_i^p$ is a p-chain on K, then

$$\varphi_p\left(\sum g_i \cdot \sigma_i^p\right) = \sum \varphi_p(g_i \cdot \sigma_i^p).$$

The sequence $\{\varphi_p\}_0^\infty$ is called the *chain mapping induced by* φ.

Theorem 3.2. *If $\varphi \colon K \to L$ is a simplicial mapping, then the sequence $\{\varphi_p\}_0^\infty$ of homomorphisms in the preceding definition is actually a chain mapping.*

PROOF. Since each φ_p is a homomorphism, then in order to show that $\partial \varphi_p = \varphi_{p-1} \partial$, it is sufficient to show that

$$\partial \varphi_p(g \cdot \sigma^p) = \varphi_{p-1} \partial (g \cdot \sigma^p)$$

for each elementary p-chain $g \cdot \sigma^p$, $p \geq 1$. Let $g \cdot \sigma^p$ be an elementary p-chain on K where $+\sigma^p = +\langle v_0 \ldots v_p \rangle$. Suppose first that φ does not collapse σ^p so that

$$\varphi_p(\sigma^p) = \langle \varphi(v_0) \ldots \varphi(v_p) \rangle.$$

Let σ_i^p be the $(p-1)$-face of σ^p obtained by deleting the ith vertex, and let $\varphi(\sigma^p)_i$ be defined in the analogous manner. Then

$$\partial \varphi_p(g \cdot \sigma^p) = \partial (g \cdot \varphi(\sigma^p)) = \sum_{i=0}^{p} (-1)^i g \cdot \varphi(\sigma^p)_i = \sum_{i=0}^{p} (-1)^i g \cdot \varphi(\sigma_i^p)$$

$$= \varphi_{p-1}\left(\sum_{i=0}^{p} (-1)^i g \cdot \sigma_i^p\right) = \varphi_{p-1} \partial (g \cdot \sigma^p).$$

Suppose that φ collapses σ^p. Without loss of generality we may assume that $\varphi(v_0) = \varphi(v_1)$. Then $\varphi_p(g \cdot \sigma^p) = 0$, so $\partial \varphi_p(g \cdot \sigma^p) = 0$, and

$$\varphi_{p-1} \partial (g \cdot \sigma^p) = \varphi_{p-1}\left(\sum_{i=0}^{p} (-1)^i g \cdot \sigma_i^p\right) = \sum_{i=0}^{p} (-1)^i \varphi_{p-1}(g \cdot \sigma_i^p).$$

For $i \geq 2$, σ_i^p contains v_0 and v_1. Since $\varphi(v_0) = \varphi(v_1)$, then φ collapses σ_i^p, $i \geq 2$, and we have

$$\varphi_{p-1} \partial (g \cdot \sigma^p) = \sum_{i=0}^{p} (-1)^i \varphi_{p-1}(g \cdot \sigma_i^p) = \varphi_{p-1}(g \cdot \sigma_0^p) - \varphi_{p-1}(g \cdot \sigma_1^p).$$

But $\sigma_0^p = \langle v_1 v_2 \ldots v_p \rangle$, $\sigma_1^p = \langle v_0 v_2 \ldots v_p \rangle$ and $\varphi(v_0) = \varphi(v_1)$ so that

$$\varphi_{p-1}(g \cdot \sigma_0^p) = \varphi_{p-1}(g \cdot \sigma_1^p).$$

Hence $\varphi_{p-1}\partial(g \cdot \sigma^p) = 0$. Thus both $\varphi_{p-1}\partial(g \cdot \sigma^p)$ and $\partial\varphi_p(g \cdot \sigma^p)$ are 0 when φ collapses σ^p. Therefore $\partial\varphi_p = \varphi_{p-1}\partial$, so $\{\varphi_p\}_0^\infty$ is a chain mapping. □

Question: The proof of Theorem 3.2 was given under the assumption that σ^p and its faces σ_i^p have orientations induced by the ordering $v_0 < v_1 < \cdots < v_p$. Why is it sufficient to consider only this orientation?

Definition. Let $|K|$ and $|L|$ be polyhedra with triangulations K and L respectively and let φ be a simplicial mapping from the vertices of K into the vertices of L. Then φ is extended to a function $\varphi: |K| \to |L|$ as follows: If $x \in |K|$, there is a simplex $\sigma^r = \langle a_0 \ldots a_r \rangle$ in K such that $x \in \sigma^r$. Then

$$x = \sum_{i=0}^{r} \lambda_i a_i$$

where the λ_i are the barycentric coordinates of x. Define

$$\varphi(x) = \sum_{i=0}^{r} \lambda_i \varphi(a_i).$$

This extended function $\varphi: |K| \to |L|$ is called a *simplicial mapping* from $|K|$ into $|L|$.

The proof of the following theorem is left as an exercise:

Theorem 3.3. *Every simplicial mapping $\varphi: |K| \to |L|$ is continuous.*

Example 3.1. Let K denote the 2-skeleton of a 3-simplex and L the closure of a 2-simplex with orientations as indicated by the arrows in Figure 3.1.

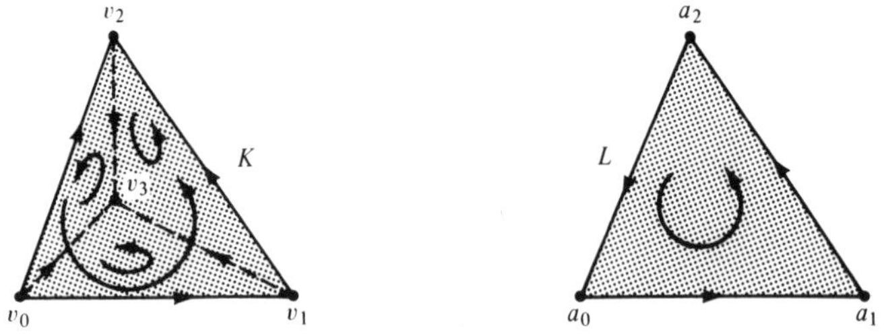

Figure 3.1

Let φ be the simplicial map from K to L defined for vertices by

$$\varphi(v_0) = \varphi(v_3) = a_0, \qquad \varphi(v_1) = a_1, \qquad \varphi(v_2) = a_2.$$

The extension process for simplicial maps determines a simplicial mapping $\varphi: |K| \to |L|$ which

(a) maps $\langle v_0v_1\rangle$, $\langle v_1v_2\rangle$, and $\langle v_2v_0\rangle$ linearly onto $\langle a_0a_1\rangle$, $\langle a_1a_2\rangle$, and $\langle a_2a_0\rangle$ respectively;
(b) maps $\langle v_1v_3\rangle$ and $\langle v_2v_3\rangle$ linearly onto $\langle a_1a_0\rangle$ and $\langle a_2a_0\rangle$ respectively;
(c) collapses $\langle v_0v_3\rangle$ to the vertex a_0;
(d) collapses $\langle v_0v_3v_2\rangle$ and $\langle v_0v_1v_3\rangle$;
(e) maps each of $\langle v_0v_1v_2\rangle$ and $\langle v_3v_1v_2\rangle$ linearly onto $\langle a_0a_1a_2\rangle$.

For the induced homomorphisms $\{\varphi_p\}$ on the chain groups we have the following:

(0) $\varphi_0: C_0(K) \to C_0(L)$ is defined by

$$\varphi_0(g\cdot\langle v_0\rangle + g_1\cdot\langle v_1\rangle + g_2\cdot\langle v_2\rangle + g_3\cdot\langle v_3\rangle) = (g_0 + g_3)\cdot\langle a_0\rangle + g_1\cdot\langle a_1\rangle + g_2\cdot\langle a_2\rangle.$$

(1) $\varphi_1: C_1(K) \to C_1(L)$ is defined by

$$\varphi_1(h_1\cdot\langle v_0v_1\rangle + h_2\cdot\langle v_1v_2\rangle + h_3\cdot\langle v_0v_2\rangle + h_4\cdot\langle v_1v_3\rangle + h_5\cdot\langle v_0v_3\rangle + h_6\cdot\langle v_2v_3\rangle) = (h_1 - h_4)\cdot\langle a_0a_1\rangle + h_2\cdot\langle a_1a_2\rangle + (h_6 - h_3)\cdot\langle a_2a_0\rangle.$$

(2) $\varphi_2: C_2(K) \to C_2(L)$ is defined by

$$\varphi_2(k_1\cdot\langle v_0v_1v_2\rangle + k_2\cdot\langle v_1v_2v_3\rangle + k_3\cdot\langle v_0v_1v_3\rangle + k_4\cdot\langle v_0v_3v_2\rangle) = (k_1 + k_2)\cdot\langle a_0a_1a_2\rangle.$$

Definition. If σ is a geometric simplex, the *open simplex* $o(\sigma)$ associated with σ consists of those points in σ all of whose barycentric coordinates are positive. If v is a vertex of a complex K, then the *star* of v, $\mathrm{st}(v)$, is the family of all simplexes σ in K of which v is a vertex. Thus $\mathrm{st}(v)$ is a subset of K. The *open star* of v, $\mathrm{ost}(v)$, is the union of all the open simplexes $o(\sigma)$ for which v is a vertex of σ. Note that $\mathrm{ost}(v)$ is a subset of the polyhedron $|K|$.

Example 3.2. If a is a vertex, $o(\langle a\rangle) = \{a\}$. For a 1-simplex $\sigma^1 = \langle a_0a_1\rangle$, $o(\sigma^1)$ is the open segment from a_0 to a_1 (not including either a_0 or a_1). For a 2-simplex σ^2, $o(\sigma^2)$ is the interior of the triangle spanned by the three vertices.

In Figure 3.2, $\mathrm{st}(v_0)$ consists of the simplexes $\langle v_0\rangle$, $\langle v_0v_1\rangle$, $\langle v_0v_2\rangle$, $\langle v_0v_3\rangle$, $\langle v_0v_4\rangle$, and $\langle v_0v_1v_2\rangle$.

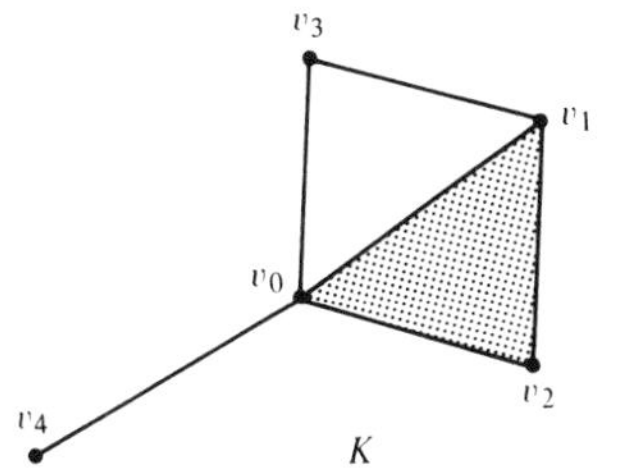

Figure 3.2

The open star of v_0, $\text{ost}(v_0)$, is the set theoretic union of $\{v_0\}$, the open segments from v_0 to v_1, v_0 to v_2, v_0 to v_3, v_0 to v_4, and the interior of $\langle v_0v_1v_2\rangle$. Note that $\text{ost}(v_0)$ is not the interior of $\text{st}(v_0)$ in any sense. The star of a vertex is a set of simplexes of K; the open star of a vertex is the union of certain point sets in the polyhedron $|K|$.

Definition. Let $|K|$ and $|L|$ be polyhedra with triangulations K and L respectively and $f: |K| \to |L|$ a continuous map. Then K is *star related to L relative to f* means that for each vertex p of K there is a vertex q of L such that

$$f(\text{ost}(p)) \subset \text{ost}(q).$$

Definition. Let X and Y be topological spaces and f, g continuous functions from X into Y. Then f is *homotopic to g* means that there is a continuous function $H: X \times [0, 1] \to Y$ from the product space $X \times [0, 1]$ into Y such that, for all $x \in X$,

$$H(x, 0) = f(x), \qquad H(x, 1) = g(x).$$

The function H is called a *homotopy* between f and g.

Note: In order to simplify notation involving homotopies, we shall use I to denote the closed unit interval $[0, 1]$.

Example 3.3. Consider the functions f and g from the unit circle S^1 into the plane given pictorially in Figure 3.3. Using the usual vector addition and scalar multiplication, a homotopy H between f and g is defined by

$$H(x, t) = (1 - t)f(x) + tg(x), \qquad x \in S^1, t \in I.$$

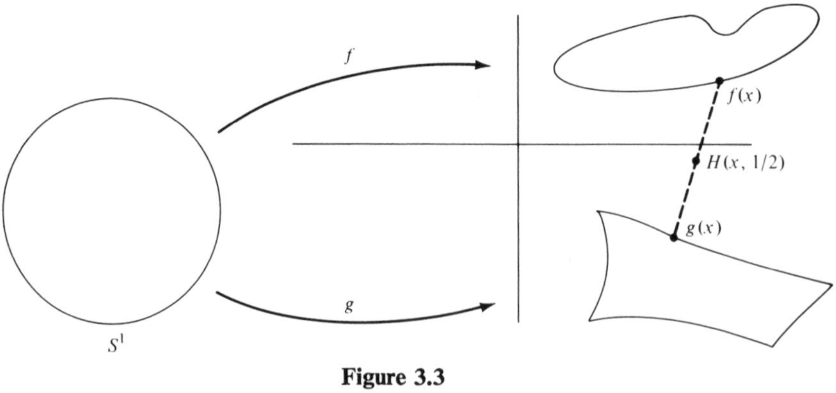

Figure 3.3

The homotopy H essentially shows how to continuously "deform" $f(x)$ into $g(x)$. Observe that if the horizontal axis were removed from the range space, then the indicated functions would not be homotopic.

Definition. Let K and L be complexes and $f: |K| \to |L|$ a continuous function. A simplicial mapping $g: |K| \to |L|$ which is homotopic to f is called a *simplicial approximation* of f.

Example 3.4. Let L be the closure of a p-simplex $\sigma^p = \langle a_0 \ldots a_p \rangle$, and let K be an arbitrary complex. Then any continuous map $f: |K| \to |L|$ has as a simplicial approximation the constant map $g: |K| \to |L|$ which collapses all of K to the vertex a_0.

As illustrated in Figure 3.4, proving that f is homotopic to g requires only the convexity of $|L|$. We define a homotopy $H: |K| \times I \to |L|$ by

$$H(x, t) = (1 - t)f(x) + ta_0, \qquad x \in |K|, t \in I.$$

Then H is continuous and

$$H(x, 0) = f(x), \qquad H(x, 1) = a_0 = g(x), \qquad x \in |K|.$$

This example illustrates one method by which homotopies will be defined in later applications.

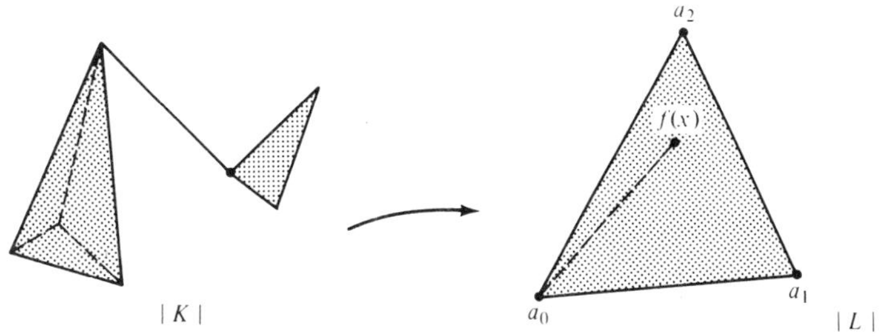

Figure 3.4

Example 3.5. Let both K and L be the 1-skeleton of the closure of a 2-simplex σ^2. Then the polyhedra $|K|$ and $|L|$ are both homeomorphic to the unit circle S^1, so we may consider any function from $|K|$ to $|L|$ as a function from S^1 to itself. For our function f, let us choose a rotation through a given angle α. Then, referring S^1 to polar coordinates, $f: S^1 \to S^1$ is defined by

$$f(1, \theta) = (1, \theta + \alpha), \qquad (1, \theta) \in S^1, 0 \leq \theta \leq 2\pi.$$

A homotopy H between f and the identity map is defined by

$$H((1, \theta), t) = (1, \theta + t\alpha), \qquad (1, \theta) \in S^1, t \in I.$$

Thus H agrees with the identity map when $t = 0$ and agrees with f when $t = 1$. At any "time" t between 0 and 1 the "t-level of the homotopy," $H(\cdot, t)$, performs a rotation of the circle through the angle $t\alpha$.

We are now ready to begin the process of replacing a continuous map $f: |K| \to |L|$ by a homotopic simplicial map g. Let us first consider the case in which K is star related to L relative to f. The following lemma will be needed; its proof is left as an exercise.

Lemma. *Vertices $v_0, \ldots, v_m$ in a complex K are vertices of a simplex of K if and only if $\bigcap_{i=0}^{m} \operatorname{ost}(v_i)$ is not empty.*

Theorem 3.4. *Let K and L be polyhedra with triangulations K and L respectively and $f: |K| \to |L|$ a continuous function such that K is star related to L relative to f. Then f has a simplicial approximation $g: |K| \to |L|$.*

PROOF. Since K is star related to L relative to f, there exists for each vertex p of K a vertex $g(p)$ of L such that

$$f(\operatorname{ost}(p)) \subset \operatorname{ost}(g(p)).$$

To see that this vertex map g is simplicial, suppose that $v_0, \ldots, v_n$ are vertices of a simplex in K. According to the lemma, this is equivalent to saying that the intersection $\bigcap_{i=0}^{n} \operatorname{ost}(v_i)$ is not empty. Hence

$$\varnothing \neq f\left(\bigcap_{i=0}^{n} \operatorname{ost}(v_i)\right) \subset \bigcap_{i=0}^{n} f(\operatorname{ost}(v_i)) \subset \bigcap_{i=0}^{n} \operatorname{ost}(g(v_i)),$$

so $\bigcap_{i=0}^{n} \operatorname{ost}(g(v_i))$ is not empty. The lemma thus insures that $g(v_0), \ldots, g(v_n)$ are vertices of a simplex in L. Then g is a simplicial vertex map and has an extension to a simplicial map $g: |K| \to |L|$.

Let $x \in |K|$ and let σ be the simplex of K of smallest dimension which contains x. Let a be any vertex of σ. Observe that $f(x) \in f(\operatorname{ost}(a))$ (why?) and that $f(\operatorname{ost}(a)) \subset \operatorname{ost}(g(a))$. Also, $g(x) \in \operatorname{ost}(g(a))$ since the barycentric coordinate of $g(x)$ with respect to $g(a)$ is greater than or equal to the (nonzero) barycentric coordinate of x with respect to a.

Let $a_0, \ldots, a_k$ denote the vertices of σ. According to the preceding paragraph, both $f(x)$ and $g(x)$ belong to $\bigcap_{i=0}^{k} \operatorname{ost}(g(a_i))$. Thus $g(a_0), \ldots, g(a_k)$ are vertices of a simplex τ in L containing both $f(x)$ and $g(x)$. Since each simplex is a convex set, then the line segment joining $f(x)$ and $g(x)$ must lie entirely in $|L|$. The map $H: |K| \times I \to |L|$ defined by

$$H(x, t) = (1 - t)f(x) + tg(x), \qquad x \in K,\ t \in I,$$

is then a homotopy between f and g, and g is a simplicial approximation of f. □

Theorem 3.4 shows that if K is star related to L relative to f, then there is a simplicial map homotopic to f. This is a big step toward our goal of replacing f by a simplicial approximation. But what if K is not star related to L relative to f? That is, what if K has some vertices $b_0, \ldots, b_n$ such that $f(\operatorname{ost}(b_i))$ is not contained in the open star of any vertex in L? We then retriangulate K systematically to produce simplexes of smaller and smaller diameters thus reducing the size of $\operatorname{ost}(b_i)$ and the size of $f(\operatorname{ost}(b_i))$ to the point that the new complex obtained from K is star related to L relative to f. This process of dividing a complex into smaller simplexes is called "barycentric subdivision." The precise definition follows.

Definition. Let $\sigma^r = \langle a_0 \ldots a_r \rangle$ be a simplex in $\mathbb{R}^n$. The point $\dot{\sigma}^r$ in σ^r all of whose barycentric coordinates with respect to $a_0, \ldots, a_r$ are equal is called the *barycenter* of σ^r. Note that if σ^0 is a 0-simplex, then $\dot{\sigma}^0$ is the vertex which determines σ^0.

The collection $\{\dot{\sigma}^k: \sigma^k \text{ is a face of } \sigma^r\}$ of all barycenters of faces of σ^r are the vertices of a complex called the *first barycentric subdivision* of $\text{Cl}(\sigma^r)$. A subset $\dot{\sigma}_0, \ldots, \dot{\sigma}_p$ of the vertices $\dot{\sigma}^k$ are the vertices of a simplex in the first barycentric subdivision provided that σ_j is a face of σ_{j+1} for $j = 0, \ldots, p - 1$.

If K is a geometric complex, the preceding process is applied to each simplex of K to produce the *first barycentric subdivision* $K^{(1)}$ of K. For $n > 1$, the nth *barycentric subdivision* $K^{(n)}$ of K is the first barycentric subdivision of $K^{(n-1)}$.

The first barycentric subdivision of K is assigned an orientation consistent with that of K as follows: Let $\langle \dot{\sigma}^0 \dot{\sigma}^1 \ldots \dot{\sigma}^p \rangle$ be a p-simplex of $K^{(1)}$ which occurs in the barycentric subdivision of a p-simplex σ^p of K. Then the vertices of $\sigma^p = \langle v_0 \ldots v_p \rangle$ may be ordered so that $\dot{\sigma}^i$ is the barycenter of $\langle v_0 \ldots v_i \rangle$ for $i = 0, \ldots, p$. We then consider $\langle \dot{\sigma}^0 \ldots \dot{\sigma}^p \rangle$ to be positively oriented if $\langle v_0 \ldots v_p \rangle$ is positively oriented and negatively oriented if $\langle v_0 \ldots v_p \rangle$ is negatively oriented. There are other simplexes of $K^{(1)}$ whose orientations are not defined by this process, and they may be oriented arbitrarily. An orientation for $K^{(1)}$ defined in this way is said to be *concordant* with the orientation of K. The same process applies inductively to higher barycentric subdivisions.

We assume in the sequel that barycentric subdivisions are concordantly oriented.

Example 3.6. Consider the complex $K = \text{Cl}(\sigma^1)$ consisting of a 1-simplex $\sigma^1 = \langle a_0 a_1 \rangle$ and two 0-simplexes $\sigma_0^0 = \langle a_0 \rangle$ and $\sigma_1^0 = \langle a_1 \rangle$. Then $\dot{\sigma}_0^0 = a_0$, $\dot{\sigma}_1^0 = a_1$, and $\dot{\sigma}^1$ is the midpoint of σ^1, as indicated in Figure 3.5. Hence the first barycentric subdivision of K has vertices a_0, a_1, and $\dot{\sigma}^1$. Since the only faces of σ^1 are $\langle a_0 \rangle$ and $\langle a_1 \rangle$, then the only 1-simplexes of $K^{(1)}$ are $\langle a_0 \dot{\sigma}^1 \rangle$ and $\langle a_1 \dot{\sigma}^1 \rangle$.

Consider σ^1 to be oriented by $a_0 < a_1$ so that $\langle a_0 a_1 \rangle$ represents the positive orientation. Then $\langle a_0 \dot{\sigma}^1 \rangle$ occurs in the subdivision of the positively oriented simplex $\langle a_0 a_1 \rangle$, and hence $\langle a_0 \dot{\sigma}^1 \rangle$ is a positively oriented simplex in $K^{(1)}$. On the other hand, $\langle a_1 \dot{\sigma}^1 \rangle$ is produced in the subdivision of the negatively oriented simplex $\langle a_1 a_0 \rangle$, so $\langle a_1 \dot{\sigma}^1 \rangle$ has negative orientation.

Figure 3.5

Example 3.7. For the complex $\text{Cl}(\sigma^2)$ in Figure 3.6(a), the barycenters of all simplexes are indicated in (b) and the first barycentric subdivision is shown in (c). The orientation for $\langle v_0v_3v_4\rangle$ is determined as follows: Vertex v_3 is the barycenter of $\langle v_0v_1\rangle$, and v_4 is the barycenter of $\langle v_0v_1v_2\rangle$. Thus, following the definition of concordant orientation, $\langle v_0v_3v_4\rangle$ is assigned positive orientation since it is produced in the subdivision of the positively oriented simplex $\langle v_0v_1v_2\rangle$. Note in Figure 3.6 that some simplexes of the barycentric subdivision are not assigned orientations by this process.

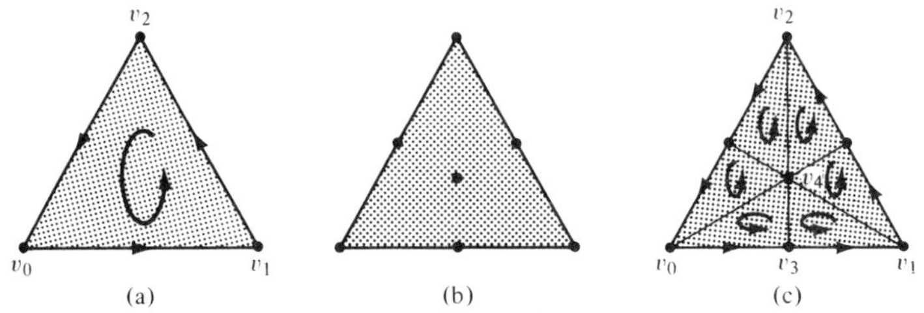

Figure 3.6

Definition. If K is a complex, the *mesh* of K is the maximum of the diameters of the simplexes of K.

It should be obvious that the mesh of the first barycentric subdivision $K^{(1)}$ of a complex K is less than the mesh of K. Hence it is reasonable to expect that the limiting value of mesh $K^{(s)}$ as s increases indefinitely is zero. Proving this requires some preliminary observations.

Let us first recall the definition of the Euclidean norm. If $x = (x_1, \ldots, x_n)$ is a point in $\mathbb{R}^n$, the *norm* of x is the number

$$\|x\| = \left\{\sum_{i=1}^{n} x_i^2\right\}^{1/2}.$$

For x, y in $\mathbb{R}^n$, the distance $d(x, y)$ from x to y is simply $\|x - y\|$. Proofs of the following facts are left as exercises:

(a) If x and y are points in a simplex σ, then there is a vertex v of σ such that

$$\|x - y\| \leq \|x - v\|.$$

(b) The diameter of a simplex of positive dimension is the length of its longest 1-face. Hence the mesh of a complex K of positive dimension is the length of its longest 1-simplex. (Any complex of dimension zero must, of course, have mesh zero.)

Theorem 3.5. *For any complex K,* $\text{limit}_{s\to\infty}$ mesh $K^{(s)} = 0$.

PROOF. Consider the first barycentric subdivision $K^{(1)}$ of K and let $\langle\dot{\sigma}\dot{\tau}\rangle$ be one of its 1-simplexes. Then σ is a face of τ. The definition of barycenter for the simplex τ insures that

$$\dot{\tau} = (1/(p + 1)) \sum_{i=0}^{p} v_i$$

where $v_0, \ldots, v_p$ are the vertices of τ.

By observation (a) above, there must be a vertex v of τ such that

$$\|\dot{\tau} - \dot{\sigma}\| \leq \|\dot{\tau} - v\|.$$

Then

$$\|\dot{\tau} - \dot{\sigma}\| \leq \|\dot{\tau} - v\| = \left\|(1/(p+1))\left(\sum_{i=0}^{p} v_i\right) - v\right\|$$

$$= \left\|(1/(p+1)) \sum_{i=0}^{p} (v_i - v)\right\| \leq (1/(p+1)) \sum_{i=0}^{p} \|v_i - v\|$$

$$\leq (p/(p+1)) \text{ mesh } K.$$

Letting n denote the dimension of K, we have $p \leq n$ so

$$\|\dot{\tau} - \dot{\sigma}\| \leq (n/(n+1)) \text{ mesh } K.$$

Since the mesh of $K^{(1)}$ is the maximum value of $\|\dot{\tau} - \dot{\sigma}\|$ for all 1-simplexes $\langle \dot{\sigma}\dot{\tau} \rangle$ in $K^{(1)}$, then

$$\text{mesh } K^{(1)} \leq (n/(n+1)) \text{ mesh } K.$$

The inductive definition of $K^{(s)}$ now insures that

$$\text{mesh } K^{(s)} \leq (n/(n+1))^s \text{ mesh } K.$$

Recalling that $\text{limit}_{s \to \infty}(n/(n+1))^s = 0$, we have the desired result. □

We are now ready for the main result of this chapter.

Theorem 3.6 (The Simplicial Approximation Theorem). *Let $|K|$ and $|L|$ be polyhedra with triangulations K and L respectively and $f\colon |K| \to |L|$ a continuous function. There is a barycentric subdivision $K^{(k)}$ of K and a continuous function $g\colon |K| \to |L|$ such that*

(a) *g is a simplicial map from $K^{(k)}$ into L, and*
(b) *g is homotopic to f.*

PROOF. We shall apply Theorem 3.4 to obtain the simplicial approximation g once an integer k for which $K^{(k)}$ is star related to L relative to f is determined. This is done using a Lebesgue number argument. Since $|L|$ is a compact metric space, the open cover $\{\text{ost}(v)\colon v \text{ is a vertex of } L\}$ has a Lebesgue number $\eta > 0$. Since f is uniformly continuous (its domain is a compact metric space), there is a positive number δ such that if $\|x - y\| < \delta$ in $|K|$, then $\|f(x) - f(y)\| < \eta$ in $|L|$. Thus, if the barycentric subdivision $K^{(k)}$ has mesh less than $\delta/2$, then $K^{(k)}$ is star related to L relative to f.

The function $g\colon |K| \to |L|$ determined by Theorem 3.4 has the required properties. □

The study of simplicial approximations to continuous functions was initiated by L. E. J. Brouwer in 1912. The Simplicial Approximation Theorem was discovered by J. W. Alexander in 1926; the proofs given above for Theorems 3.4 and 3.6 are essentially his original ones [27].

After a long, difficult sequence of proofs, it may be comforting to know that the *existence* of simplicial approximations is the important thing. We will not have to perform tedious constructions of simplicial approximations; any simplicial approximation of the type guaranteed by the Simplicial Approximation Theorem will usually do quite nicely.

3.3 Induced Homomorphisms on the Homology Groups

Definition. Let $|K|$ and $|L|$ be polyhedra with triangulations K and L respectively and $f: |K| \to |L|$ a continuous map. By the Simplicial Approximation Theorem, there is a barycentric subdivision $K^{(k)}$ of K and a simplicial mapping $g: |K| \to |L|$ which is homotopic to f. Theorems 3.1 and 3.2 insure that g induces homomorphisms $g_p^*: H_p(K) \to H_p(L)$ in each dimension p. This sequence of homomorphisms $\{g_p^*\}$ is called the *sequence of homomorphisms induced by* f.

The preceding definition raises a question about the uniqueness of the sequence of homomorphisms induced by f. It can be shown, however, that the sequence $\{g_p^*\}$ is unique and, in particular, does not depend on the admissible choices for the degree k of the barycentric subdivision or on the admissible choices for the simplicial map g. The sequence is thus usually written $\{f_p^*\}$ instead of $\{g_p^*\}$ since it is completely determined by f. Showing that the sequence is unique requires some concepts that we have not yet developed. The proof will therefore be postponed until Section 1 of Chapter 7. Those who cannot wait to see the proof may read that section now.

We shall illustrate the utility of induced homomorphisms by proving that two Euclidean spaces of different dimensions are not homeomorphic. This was first proved by L. E. J. Brouwer in 1911; it is, of course, not a surprising result. Any reader who feels that this is a trivial application, however, is invited to produce his own proof before reading further.

The following lemma is left as an exercise:

Lemma. *If $f: |K| \to |L|$ and $h: |L| \to |M|$ are continuous maps on the indicated polyhedra, then $(hf)_p^*: H_p(K) \to H_p(M)$ is the composition*

$$h_p^* f_p^*: H_p(K) \to H_p(M)$$

in each dimension p.

Theorem 3.7 (Invariance of Dimension). *If $m \neq n$, then*

(a) *S^m and S^n are not homeomorphic, and*
(b) *$\mathbb{R}^m$ and $\mathbb{R}^n$ are not homeomorphic.*

Proof. (a) Suppose to the contrary that there is a homeomorphism $h: S^m \to S^n$ from S^m onto S^n with inverse $h^{-1}: S^n \to S^m$. Then $h^{-1}h$ and hh^{-1} are the identity maps on S^m and S^n respectively. Note that the identity map i on a polyhedron $|K|$ induces the identity isomorphism $i_p^*: H_p(K) \to H_p(K)$ in each dimension p. Then

$$(hh^{-1})_p^* = h_p^* h_p^{-1*}: H_p(S^n) \to H_p(S^n),$$
$$(h^{-1}h)_p^* = h_p^{-1*} h_p^*: H_p(S^m) \to H_p(S^m)$$

are identity isomorphisms in each dimension, so h_p^* is an isomorphism between $H_p(S^m)$ and $H_p(S^n)$. Comparison of homology groups (Theorem 2.9) reveals that this is impossible since $m \neq n$. Hence S^m and S^n are not homeomorphic when $m \neq n$.

(b) Recall from point-set topology that S^n is the one point compactification of $\mathbb{R}^n$. Thus if $\mathbb{R}^m$ and $\mathbb{R}^n$ are homeomorphic, it must be true that their one point compactifications S^m and S^n are homeomorphic too. This contradicts part (a) if $m \neq n$. □

A special case of the definition of induced homomorphisms for maps on spheres will be of particular importance.

Definition. Let $f: S^n \to S^n$, $n \geq 1$, be a continuous function from the n-sphere into itself. Let K be a triangulation of S^n. Since K is an orientable n-pseudomanifold, Theorem 2.10 and its proof show that it is possible to orient K so that the n-chain

$$z_n = \sum_{\sigma^n \in K} 1 \cdot \sigma^n$$

is an n-cycle whose homology class $[z_n]$ is a generator of the infinite cyclic group $H_n(K)$. This homology class is called a *fundamental class*. If $f_n^*: H_n(K) \to H_n(K)$ is the homomorphism in dimension n induced by f, then there is an integer ρ such that

$$f_n^*([z_n]) = \rho[z_n].$$

The integer ρ is called the *degree* of the map f and is denoted $\deg(f)$.

The degree of a map on S^n was originally defined by L. E. J. Brouwer. The above definition is a modern version equivalent to his original one which is stated here for comparison. The student should feel free to use whichever definition fits best in a particular situation since they are equivalent.

Alternate Definition. Suppose that $f: S^n \to S^n$ is a continuous map and S^n is triangulated by a complex K. Choose a barycentric subdivision $K^{(k)}$ of K for which there is a simplicial mapping $\varphi: |K^{(k)}| \to |K|$ homotopic to f. Let τ be any positively oriented n-simplex in K. Let p be the number of positively oriented n-simplexes σ in $K^{(k)}$ such that $\varphi(1 \cdot \sigma) = 1 \cdot \tau$, and let q be the number of positively oriented n-simplexes μ in $K^{(k)}$ such that $\varphi(1 \cdot \mu) = -1 \cdot \tau$. Then the integer $p - q$ is independent of the choice of τ (the same integer $p - q$ results for each n-simplex of K) and is called the *degree* of the map f.

In Brouwer's definition it can be shown that the degree of f is independent of the admissible choices for K, $K^{(k)}$, and φ (see, for example, [9], section 6-14).

Intuitively, the definition states that the degree of a map $f: S^n \to S^n$ is the number of times that f "wraps the domain around the range."

Theorem 3.8. (a) *If $f: S^n \to S^n$ and $g: S^n \to S^n$ are continuous maps, then* $\deg(gf) = \deg(g)\cdot\deg(f)$.
(b) *The identity map $i: S^n \to S^n$ has degree $+1$.*
(c) *A homeomorphism $h: S^n \to S^n$ has degree ± 1.*

PROOF. (a) Choose a triangulation K of S^n with fundamental class $[z_n]$ and consider the induced homomorphisms

$$f_n^*: H_n(K) \to H_n(K), \qquad g_n^*: H_n(K) \to H_n(K).$$

Then

$$(gf)_n^*([z_n]) = \deg(gf)\cdot[z_n],$$
$$g_n^* f_n^*([z_n]) = g_n^*(\deg(f)\cdot[z_n]) = \deg(g)\cdot\deg(f)\cdot[z_n].$$

Since the lemma preceding Theorem 3.7 insures that $(gf)_n^* = g_n^* f_n^*$, then $\deg(gf) = \deg(g)\cdot\deg(f)$.

(b) In Brouwer's definition of degree, it is obvious that for the identity map i, $p = 1$ and $q = 0$ so $\deg(i) = 1 - 0 = 1$.

(c) Letting h^{-1} denote the inverse of h, we have

$$1 = \deg(i) = \deg(hh^{-1}) = \deg(h)\deg(h^{-1}).$$

Since $\deg(h)$ must be an integer, then $\deg(h) = \pm 1$. It also follows that h and h^{-1} have the same degree. □

The following theorem was proved by Brouwer in 1912:

Theorem 3.9 (Brouwer's Degree Theorem). *If two continuous maps $f, g: S^n \to S^n$ are homotopic, then they have the same degree.*

PROOF. Let K be a triangulation of S^n and let $h: S^n \times I \to S^n$ be a homotopy such that

$$h(x, 0) = f(x), \qquad h(x, 1) = g(x), \qquad x \in S^n.$$

For convenience in notation we let h_t denote the restriction of h to $S^n \times \{t\}$. Thus $h_0 = f$ and $h_1 = g$.

Let ϵ be a Lebesque number for the open cover $\{\text{ost}(w_i): w_i \text{ is a vertex of } K\}$. Since h is uniformly continuous, there is a positive number δ such that if A and B are subsets of S^n and I respectively with diameters $\text{diam}(A) < \delta$ and $\text{diam}(B) < \delta$, then $\text{diam}(h(A \times B)) < \epsilon$. Let $K^{(k)}$ be a barycentric subdivision of K of mesh less than $\delta/2$ so that if v is a vertex of $K^{(k)}$, then $\text{diam}(\text{ost}(v)) < \delta$. Let

$$0 = t_0 < t_1 < \cdots < t_q = 1$$

be a partition of I for which successive points differ by less than δ. Then each set $h(\text{ost}(v_i) \times [t_{j-1}, t_j])$, v_i a vertex of $K^{(k)}$ and t_{j-1}, t_j successive members of the partition, has diameter less than ϵ and is therefore contained in $\text{ost}(w_{ij})$ for some vertex w_{ij} of K.

Thus if $t_{j-1} \le t \le t_j$, the value of the simplicial map φ_t approximating h_t given by the Simplicial Approximation Theorem may be defined by letting $\varphi_t(v_i) = w_{ij}$. We therefore conclude that all the maps h_t for $t_{j-1} \le t \le t_j$ have the same degree. Since any two successive intervals $[t_{j-1}, t_j]$ and $[t_j, t_{j+1}]$ have t_j in common, it follows that the degree of h_t is constant for $0 \le t \le 1$. In particular, $h_0 = f$ and $h_1 = g$ have the same degree. □

The preceding method of proof can be extended to show that homotopic maps from one polyhedron to another induce identical sequences of homomorphisms on the homology groups. Along with the preceding theorem, Brouwer proved a partial converse: If f and g are continuous maps on the 2-sphere which have the same degree, then they are homotopic. This conclusion was extended to arbitrary dimension by Heinz Hopf (1894–1971) in 1927. The combined results form the famous Hopf Classification Theorem, which is stated here without proof:

Theorem 3.10 (The Hopf Classification Theorem). *Two continuous maps f, g from S^n to S^n are homotopic if and only if they have the same degree.*

Hopf extended Brouwer's definition of degree to maps from polyhedra into spheres and, in 1933, extended his classification theorem to such maps: *If X is a polyhedron of dimension not exceeding n, then two maps f and g from X into S^n are homotopic if and only if they have the same degree.* Proofs can be found in [20] and in Hopf's original paper [41].

3.4 The Brouwer Fixed Point Theorem and Related Results

Definition. If $f: X \to X$ is a continuous function from a space X into itself, then a point x_0 in X is a *fixed point* of f means that $f(x_0) = x_0$.

Theorems about fixed points have far reaching applications in mathematics. The existence of a solution for a differential or integral equation, for example, is often equivalent to the existence of a fixed point of a linear operator on a function space. (In this connection see Picard's Theorem from differential equations.) In this section we shall prove the classic fixed point theorem of L. E. J. Brouwer and some related results about S^n.

Definition. A continuous function $g: X \to Y$ from a space X into a space Y which is homotopic to a constant map is said to be *null-homotopic* or *inessential.*

Definition A space X is *contractible* means that the identity function $i: X \to X$ is null-homotopic. In other words, X is contractible if there is a point x_0 in X and a homotopy $H: X \times I \to X$ such that

$$H(x, 0) = x, \qquad H(x, 1) = x_0, \qquad x \in X.$$

The homotopy H is called a *contraction* of the space X.

Example 3.8. The unit disk $D = \{x = (x_1, x_2) \in \mathbb{R}^2 : x_1^2 + x_2^2 \leq 1\}$ is contractible. We let $x_0 = (0, 0)$ be the origin and define a contraction by

$$H((x_1, x_2), t) = ((1 - t)x_1, (1 - t)x_2), \qquad (x_1, x_2) \in D,\ t \in I.$$

Imagining the disk as a sheet of rubber, the contraction essentially "squeezes" the disk to a single point.

This intuitive idea of contractibility suggests that the circle is not contractible. This is in fact true and is a consequence of the following theorem of L. E. J. Brouwer.

Theorem 3.11 *The n-sphere S^n is not contractible for any $n \geq 0$.*

PROOF. The identity map on S^n has degree 1 for $n \geq 1$, and any constant map has degree 0. Since homotopic maps have the same degree (Theorem 3.9), then the identity is not null-homotopic, and S^n is not contractible for $n \geq 1$.

For the case $n = 0$, we observe that

$$S^0 = \{x \in \mathbb{R} : x^2 = 1\} = \{-1, 1\}$$

is a discrete space and therefore not contractible. □

Theorem 3.12 (The Brouwer No Retraction Theorem). *There does not exist a continuous function from the $(n + 1)$-ball*

$$B^{n+1} = \left\{x = (x_1, x_2, \ldots, x_{n+1}) \in \mathbb{R}^{n+1} : \sum x_i^2 \leq 1\right\}$$

onto S^n which leaves each point of S^n fixed, $n \geq 0$.

PROOF. Assuming that a map $f: B^{n+1} \to S^n$ such that $f(x) = x$ for each x in S^n does exist, define a homotopy

$$H: S^n \times I \to S^n$$

by

$$H(x, t) = f((1 - t)x), \qquad x \in S^n,\ t \in I.$$

Here $(1 - t)x$ denotes the usual scalar product (real number multiplied by a vector) in $\mathbb{R}^n$. Then H is a contraction on S^n contradicting Theorem 3.11. Thus no such map f exists. □

Theorem 3.13 (The Brouwer Fixed Point Theorem). *If $f: B^{n+1} \to B^{n+1}$ is continuous map from the $(n + 1)$-ball into itself and $n \geq 0$, then f has at least one fixed point.*

PROOF. Suppose on the contrary that f has no fixed point. Then for each $x \in B^{n+1}$, $f(x)$ and x are distinct points. For any x consider the half-line from $f(x)$ through x, and let $g(x)$ denote the intersection of this ray with S^n, as shown in Figure 3.7.

Then $g: B^{n+1} \to S^n$ is continuous, and $g(x) = x$ for each $x \in S^n$. This contradicts the preceding theorem, so we conclude that the assumption that f has no fixed point must be false. □

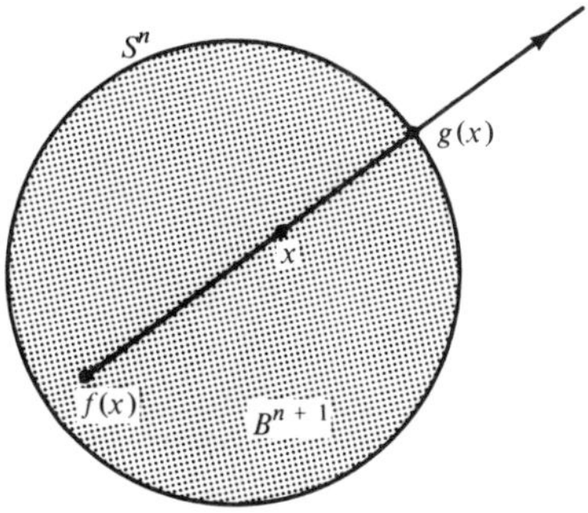

Figure 3.7

The Brouwer Fixed Point Theorem was first proved by Brouwer in 1912. The proof given in the text is not his original one.

Definition. For each integer i with $1 \leq i \leq n + 1$, the map

$$r_i: S^n \to S^n$$

defined by

$$r_i(x_1, x_2, \ldots, x_{n+1}) = (x_1, \ldots, x_{i-1}, -x_i, x_{i+1}, \ldots, x_{n+1}), \qquad (x_1, \ldots, x_{n+1}) \in S^n,$$

(with obvious modifications when $i = 1$ or $n + 1$) is called the *reflection of S^n with respect to the x_i axis.*

Definition. The map $r: S^n \to S^n$ defined by

$$r(x) = -x, \qquad x \in S^n,$$

is called the *antipodal map* on S^n.

For $x = (x_1, x_2, \ldots, x_{n+1}) \in S^n$, $r_i(x)$ and x differ only in the ith coordinate, and the ith coordinate of $r_i(x)$ is the negative of the ith coordinate of x. The antipodal map r takes each point x in S^n to the diametrically opposite point $r(x) = -x$ each of whose coordinates is the negative of the corresponding coordinate of x. It should be clear that the antipodal map r is the composition $r_1 r_2 \ldots r_{n+1}$ of the reflections of S^n in the respective axes. The proof of the following lemma is left as an exercise.

Lemma. (a) *Each reflection r_i on S^n has degree* -1.
(b) *The antipodal map on S^n has degree* $(-1)^{n+1}$.

Definition. A *continuous unit tangent vector field*, or simply *vector field*, on S^n is a continuous function $f: S^n \to S^n$ such that x and $f(x)$ are perpendicular for each x in S^n.

In order to get an intuitive picture of a vector field, let us first review the concept of perpendicular vectors. Recall from sophomore Calculus that two vectors $x = (x_1, x_2)$ and $y = (y_1, y_2)$ in the plane are perpendicular if and only if their dot product (or inner product)

$$x \cdot y = x_1 y_1 + x_2 y_2 = 0.$$

Perpendicularity is extended to vectors of higher dimension by the following definition: Two vectors $x = (x_1, \ldots, x_n)$ and $y = (y_1, \ldots, y_n)$ in $\mathbb{R}^n$ are *perpendicular* if and only if their dot product (Appendix 2)

$$x \cdot y = x_1 y_1 + x_2 y_2 + \cdots + x_n y_n = 0.$$

A vector field f on S^n is then interpreted as follows: f is a continuous function which associates with vector x of unit length in $\mathbb{R}^{n+1}$ a unit vector $f(x)$ in $\mathbb{R}^{n+1}$ such that x and $f(x)$ are perpendicular. If we imagine that $f(x)$ is transposed so that it begins at point x on S^n, then $f(x)$ must be tangent to the sphere S^n. This idea is illustrated in Figure 3.8.

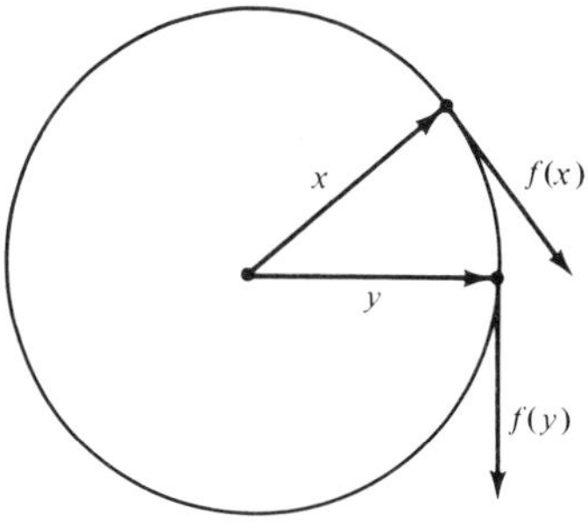

Figure 3.8

It should be clear that the following scheme describes a vector field on S^1. For each x in S^1, let $f(x)$ denote a vector of unit length beginning at point x and pointing in the clockwise direction tangent to S^1. Having all vectors $f(x)$ point in the counterclockwise direction also produces a vector field on S^1. The requirement of continuity for f rules out the possibility of having $f(x)$ in the clockwise direction for some values of x and in the counterclockwise direction for others.

Theorem 3.14 (The Brouwer–Poincaré Theorem). *There is a vector field on S^n, $n \geq 1$, if and only if n is odd.*

Proof. If n is odd, a vector field f on S^n can be defined by

$$f(x_1, x_2, \ldots, x_{n+1}) = (x_2, -x_1, x_4, -x_3, \ldots, x_{n+1}, -x_n), \qquad (x_1, x_2, \ldots, x_{n+1}) \in S^n.$$

It is clear that f is a continuous function from S^n into S^n. The proof that f is a vector field is completed by observing that, for each x in S^n,

$$x \cdot f(x) = (x_1 x_2 - x_1 x_2) + (x_3 x_4 - x_3 x_4) + \cdots + (x_n x_{n+1} - x_n x_{n+1}) = 0.$$

Suppose now that $g: S^n \to S^n$ is a vector field where n is an even integer. This assumption will lead to a contradiction. Define a homotopy $h: S^n \times I \to S^n$ by

$$h(x, t) = x\cos(t\pi) + g(x)\sin(t\pi), \qquad x \in S^n,\ t \in I.$$

Then

$$\begin{aligned}\|h(x, t)\|^2 &= h(x, t)\cdot h(x, t)\\ &= \|x\|^2\cos^2(t\pi) + 2x\cdot g(x)\cos(t\pi)\sin(t\pi) + \|g(x)\|^2\sin^2(t\pi)\\ &= 1^2\cos^2(t\pi) + (2)(0)\cos(t\pi)\sin(t\pi) + 1^2\sin^2(t\pi)\\ &= 1,\end{aligned}$$

so h is a homotopy on S^n. But

$$h(x, 0) = x, \qquad h(x, 1) = -x, \qquad x \in S^n,$$

so h is a homotopy between the identity map and the antipodal map on S^n. However, the identity map has degree 1 and the antipodal map has degree $(-1)^{n+1} = -1$ since n is even. This contradicts Brouwer's Theorem on the degree of homotopic maps (Theorem 3.9). Thus S^n has a vector field if and only if n is odd. □

The main part of the Brouwer–Poincaré Theorem (there is no vector field on a sphere of even dimension) was conjectured by Poincaré and first proved by Brouwer. For $n = 2$, the result can be visualized as follows: Imagine a 2-sphere with a unit vector emanating from each point; think of each vector as a hair. Finding a vector field for S^2 is equivalent to describing a method for "combing the hairs" so that each one is tangent to the sphere and so that their directions vary continuously. In other words, there must be no parts or whorls in the hairs. According to the Brouwer–Poincaré Theorem, such a hairstyle is impossible for spheres of even dimension. Because of this analogy, the theorem is sometimes called the "Tennis Ball Theorem."

Exercises

1. Give an example of two polyhedra $|K|$ and $|L|$ for which $H_p(K)$ and $H_p(L)$ are isomorphic for each value of p, but $|K|$ and $|L|$ are not homeomorphic.

2. Verify in the proof of Theorem 3.1 that φ_p^* is a homomorphism. Show in particular that if $[z_p] = [w_p]$ in $H_p(K)$, then $[\varphi_p(z_p)] = [\varphi_p(w_p)]$ in $H_p(L)$.

3. Prove Theorem 3.3.

4. (a) For the simplicial map φ of Example 3.1, describe the induced homomorphisms $\varphi_p^*: H_p(K) \to H_p(L)$.
   (b) Prove that if L is replaced by its 1-skeleton, then the map f is not simplicial.

5. Choose triangulations for the 2-sphere S^2 and torus T, and let $\varphi: S^2 \to T$ be a simplicial map. Prove that the induced homomorphism $\varphi_p^*: H_p(S^2) \to H_p(T)$ is trivial for $p \geq 1$. Show that this result does not hold if the roles of S^2 and T are interchanged.

6. Let X and Y be topological spaces and let M denote the set of all continuous maps f from X into Y. For brevity let us agree that $f \sim g$ means that f is homotopic to g. Prove that $\sim$ is an equivalence relation on M.

7. (a) Prove that every convex subset of $\mathbb{R}^n$ is contractible.
   (b) Given that Y is contractible, prove that every continuous function from a space X into Y is null-homotopic.

8. Prove that vertices $v_0, v_1, \ldots, v_m$ of a complex K are vertices of a simplex in K if and only if $\bigcap_{i=0}^{m} \text{ost}(v_i)$ is not empty.

9. Prove the following facts:
   (a) If x and y are points in a simplex σ, then there is a vertex v of σ such that $\|x - y\| \leq \|x - v\|$.
   (b) The diameter of a simplex σ^p, $p \geq 1$, is the maximum length of its 1-faces.
   (c) The mesh of a complex K is the maximum length of its 1-simplexes if K has positive dimension.

10. Answer the following questions about the proof of Theorem 3.4:
    (a) If σ is the simplex of smallest dimension in K containing a given point x, why is x in $\text{ost}(a_i)$ for each vertex a_i of σ?
    (b) Why is the function H continuous?

11. Complete the details in the proof of Theorem 3.6 by proving the following:
    (a) If v is a vertex of K, then the diameter of $\text{ost}(v)$ does not exceed twice the mesh of K.
    (b) If v is a vertex of K, then $\text{ost}(v)$ is an open subset of $|K|$. (Recall that $|K|$ has the Euclidean subspace topology.)
    (c) Prove that every polyhedron is a compact metric space.
    (d) Show that the function g in the proof of Theorem 3.6 has the required properties.

12. Prove that the antipodal map on S^n has degree $(-1)^{n+1}$.

13. (a) Prove the lemma preceding Theorem 3.7: If $f\colon |K| \to |L|$ and $h\colon |L| \to |M|$ are continuous maps, then $(hf)_p^* = h_p^* f_p^*$ in each dimension p.
    (b) Prove that if two polyhedra $|K|$ and $|L|$ are homeomorphic, then $H_p(K) \cong H_p(L)$ in each dimension p.

14. Prove the following fact about maps $f, g\colon S^n \to S^n$: If $\deg(f) = \deg(g)$, then $g_n^* = f_n^*\colon H_n(S^n) \to H_n(S^n)$.

15. Prove that a discrete space X is contractible if and only if X has only one point.

16. Is every subspace of a contractible space contractible? Explain.

17. Show that if $|K|$ is contractible, then $H_p(K) = \{0\}$ for $p \geq 1$ and $H_0(K) \cong \mathbb{Z}$.

18. In the text the Brouwer Fixed Point Theorem was proved as a consequence of the Brouwer No Retraction Theorem. Reverse this process to show that the Fixed Point Theorem implies the No Retraction Theorem.

19. **Definition.** Let X be a topological space and B a subspace of X. If there is a continuous map $f\colon X \to B$ which leaves each point of B fixed, then B is called a *retract* of X. The function f is a *retraction* of X onto B.

Let A and K be complexes for which A is a subset of K and $|A|$ is a retract of $|K|$. Prove that $H_p(K)$ has a subgroup isomorphic to $H_p(A)$ in each dimension p.

20. Prove the Brouwer No Retraction Theorem by comparing the homology groups of S^n and B^{n+1}. (*Hint*: Assuming that there is a retraction $f: B^{n+1} \to S^n$, let $i: S^n \to B^{n+1}$ denote the inclusion map. Then $fi: S^n \to S^n$ is the identity map. Consider the homomorphism induced on $H_n(S^n)$.

21. Let f, g be continuous maps from a space X into S^n such that $f(x)$ and $g(x)$ are never antipodal points, i.e., $f(x) = -g(x)$ for no x. Prove that f and g are homotopic.

22. Find an explicit formula for the vector field on S^1 which produces tangent vectors with the clockwise orientation. Repeat for the counterclockwise orientation.

23. Prove that every vector field on S^n (n odd) is homotopic to the identity map and to the antipodal map.

24. Let n be an even positive integer and $f: S^n \to E^{n+1}$ a continuous map such that x and $f(x)$ are perpendicular for each $x \in S^n$. Prove that there is a point x in S^n for which $f(x) = 0$.

25. Consider the circle S^1 with multiplication given by the complex numbers. Prove that the map $f(x) = x^n$, n a positive integer, has degree n. What is the degree of the map $g(x) = 1/x$?

26. Let $g: S^n \to S^n$ be a continuous map for which the range is a proper subset of S^n. Prove that g is null-homotopic and that $\deg(g) = 0$.

27. (a) Let $g: S^n \to S^n$ be a continuous map for which there is a continuous extension $G: B^{n+1} \to S^n$. Prove that g is null-homotopic.

(b) Prove the converse: If $g: S^n \to S^n$ is null-homotopic, then g has a continuous extension $G: B^{n+1} \to S^n$. (*Hint*: B^{n+1} can be considered to be the quotient space of $S^n \times [0, 1]$ obtained by identifying $S^n \times \{1\}$ to a single point.)

28. Let K, L, and M be complexes and $f: |K| \to |L|$ and $g: |L| \to |M|$ continuous functions. If K is star related to L relative to f and L is star related to M relative to g, prove that K is star related to M relative to gf.

29. Show that every continuous function $f: |K| \to |L|$ from a polyhedron $|K|$ to a polyhedron $|L|$ can be arbitrarily approximated in terms of distance by a simplicial approximation. More precisely, prove the following:

Theorem. *Let $f: |K| \to |L|$ be a continuous map on the indicated polyhedra and ϵ a positive number. There are barycentric subdivisions $K^{(i)}$ and $L^{(j)}$ and a continuous map $g: |K| \to |L|$ such that*

(a) g is a simplicial map with respect to $K^{(i)}$ and $L^{(j)}$,

(b) g is homotopic to f, and

(c) the distance $\|f(x) - g(x)\|$ is less than ϵ for all x in $|K|$.

30. (a) Prove that every barycentric subdivision of an n-pseudomanifold is an n-pseudomanifold.

(b) If K is an orientable pseudomanifold, is each barycentric subdivision of K orientable? Prove that your answer is correct.

(c) Repeat part (b) for the nonorientable case.

4 The Fundamental Group

4.1 Introduction

We turn now to the investigation of the structure of a topological space by means of paths or curves in the space. Recall that in Chapter 1 we decided that two closed paths in a space are homotopic provided that each of them can be "continuously deformed into the other." In Figure 4.1, for example, paths C_2 and C_3 are homotopic to each other and C_1 is homotopic to a constant path. Path C_1 is not homotopic to either C_2 or C_3 since neither C_2 nor C_3 can be pulled across the hole that they enclose.

In this chapter we shall make precise this intuitive idea of homotopic paths. The basic idea is a special case of the homotopy relation for continuous functions which we considered in the proof of the Simplicial Approximation Theorem (Theorem 3.6).

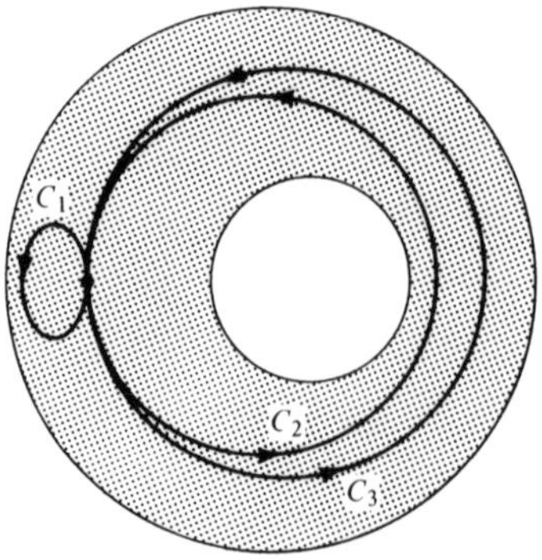

Figure 4.1

4.2 Homotopic Paths and the Fundamental Group

Definition. A *path* in a topological space X is a continuous function α from the closed unit interval $I = [0, 1]$ into X. The points $\alpha(0)$ and $\alpha(1)$ are the *initial point* and *terminal point* of α respectively. Paths α and β with common initial point $\alpha(0) = \beta(0)$ and common terminal point $\alpha(1) = \beta(1)$ are *equivalent* provided that there is a continuous function $H: I \times I \to X$ such that

$$H(t, 0) = \alpha(t), \qquad H(t, 1) = \beta(t), \qquad t \in I,$$
$$H(0, s) = \alpha(0) = \beta(0), \qquad H(1, s) = \alpha(1) = \beta(1), \qquad s \in I.$$

The function H is called a *homotopy* between α and β. For a given value of s, the restriction of H to $I \times \{s\}$ is called the *s-level* of the homotopy and is denoted $H(\cdot, s)$.

Definition. A *loop* in a topological space X is a path α in X with $\alpha(0) = \alpha(1)$. The common value of the initial point and terminal point is referred to as the *base point* of the loop. Two loops α and β having common base point x_0 are *equivalent* or *homotopic modulo* x_0 provided that they are equivalent as paths. In other words, α and β are homotopic modulo x_0 (denoted $\alpha \sim_{x_0} \beta$) provided that there is a homotopy $H: I \times I \to X$ such that

$$H(\cdot, 0) = \alpha, \qquad H(\cdot, 1) = \beta, \qquad H(0, s) = H(1, s) = x_0, \qquad s \in I.$$

Since $H(0, s)$ and $H(1, s)$ always have value x_0 regardless of the choice of s in $[0, 1]$, it is sometimes said that the base point "stays fixed throughout the homotopy."

Example 4.1. The paths α and β in Figure 4.2 are equivalent. A homotopy H demonstrating the equivalence is defined by

$$H(t, s) = s\beta(t) + (1 - s)\alpha(t), \qquad (s, t) \in I \times I.$$

The homotopy essentially "pulls α across to β" without disturbing the end points. If the space had a "hole" between the ranges of α and β, then the paths would not be equivalent.

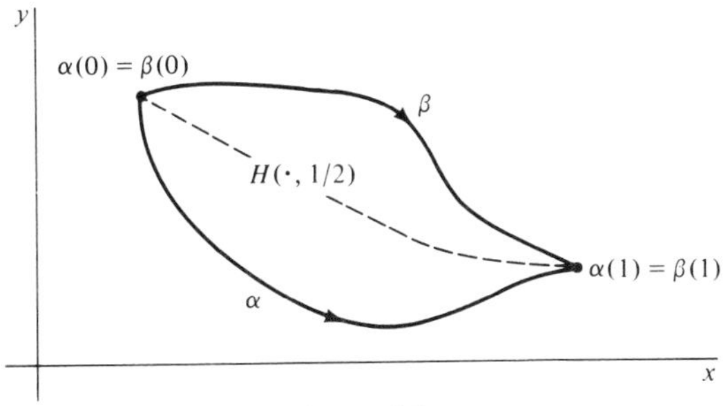

Figure 4.2

The following lemma from point-set topology will be used repeatedly in this chapter. Its proof is left as an exercise.

The Continuity Lemma. *Let X be a topological space with closed subsets A and B such that $A \cup B = X$. Let $f: A \to Y$ and $g: B \to Y$ be continuous maps to a space Y such that $f(x) = g(x)$ for each x in $A \cap B$. Then the map $h: X \to Y$ defined by*

$$h(x) = \begin{cases} f(x) & \text{if } x \in A \\ g(x) & \text{if } x \in B \end{cases}$$

is continuous.

Theorem 4.1. (a) *Equivalence of paths is an equivalence relation on the set of paths in a space X.*

(b) *Equivalence of loops is an equivalence relation on the set of loops in X with base point x_0.*

PROOF. We shall prove (b) and leave to the reader the obvious modifications needed for a proof of (a).

Consider the set of loops in X having base point x_0. Any such loop α is equivalent to itself under the homotopy

$$F(t, s) = \alpha(t), \qquad (t, s) \in I \times I.$$

Thus the relation $\sim_{x_0}$ is reflexive.

Suppose $\alpha \sim_{x_0} \beta$. Then there is a homotopy $H: I \times I \to X$ satisfying

$$H(\cdot, 0) = \alpha, \qquad H(\cdot, 1) = \beta, \qquad H(0, s) = H(1, s) = x_0, \qquad s \in I.$$

Then the homotopy

$$\bar{H}(t, s) = H(t, 1 - s), \qquad (s, t) \in I \times I,$$

shows that $\beta \sim_{x_0} \alpha$ and hence that equivalence of loops is a symmetric relation.

Suppose now that for the loops α, β, and γ we have $\alpha \sim_{x_0} \beta$ and $\beta \sim_{x_0} \gamma$. Then there are homotopies H and K such that

$$H(\cdot, 0) = \alpha, \qquad H(\cdot, 1) = \beta, \qquad H(0, s) = H(1, s) = x_0, \qquad s \in I,$$
$$K(\cdot, 0) = \beta, \qquad K(\cdot, 1) = \gamma, \qquad K(0, s) = K(1, s) = x_0, \qquad s \in I.$$

The required homotopy L between α and γ is defined by

$$L(t, s) = \begin{cases} H(t, 2s) & \text{if } 0 \le s \le \frac{1}{2} \\ K(t, 2s - 1) & \text{if } \frac{1}{2} \le s \le 1. \end{cases}$$

The continuity of L follows from the Continuity Lemma with $A = I \times [0, \frac{1}{2}]$ and $B = I \times [\frac{1}{2}, 1]$. Thus $\alpha \sim_{x_0} \gamma$, so $\sim_{x_0}$ is an equivalence relation. □

Definition. If α and β are paths in X with $\alpha(1) = \beta(0)$, then the *path product* $\alpha * \beta$ is the path defined by

$$\alpha * \beta(t) = \begin{cases} \alpha(2t) & \text{if } 0 \leq t \leq \frac{1}{2} \\ \beta(2t - 1) & \text{if } \frac{1}{2} \leq t \leq 1. \end{cases}$$

The continuity of $\alpha * \beta$ is an immediate consequence of the Continuity Lemma.

Thinking of the variable t as time, a path α in X can be visualized by the motion of a point beginning at $\alpha(0)$ and tracing a continuous route to $\alpha(1)$. A product $\alpha * \beta$ is then visualized as follows: The moving point begins at $\alpha(0)$ and follows path α at twice the normal rate, arriving at $\alpha(1)$ when $t = \frac{1}{2}$. The point then follows path β at twice the normal rate and arrives at $\beta(1)$ at time $t = 1$. Note that the condition $\alpha(1) = \beta(0)$ is required for the product of paths in order to avoid discontinuities.

We shall be primarily concerned with products of loops α and β having common base point x_0. In this case the product $\alpha * \beta$ is also a loop with base point x_0. The following lemma is left as an exercise:

Lemma. *Suppose that loops α, α', β, β' in a space X all have base point x_0 and satisfy the relations $\alpha \sim_{x_0} \alpha'$ and $\beta \sim_{x_0} \beta'$. Then the products $\alpha * \beta$ and $\alpha' * \beta'$ are homotopic modulo x_0.*

Definition. Consider the family of loops in X with base point x_0. Homotopy modulo x_0 is an equivalence relation on this family and therefore partitions it into disjoint equivalence classes, $[\alpha]$ denoting the equivalence class determined by loop α. The class $[\alpha]$ is called the *homotopy class* of α. The set of such homotopy classes is denoted by $\pi_1(X, x_0)$. If $[\alpha]$ and $[\beta]$ belong to $\pi_1(X, x_0)$, then the *product* $[\alpha] \circ [\beta]$ is defined as follows:

$$[\alpha] \circ [\beta] = [\alpha * \beta].$$

Thus the product of two homotopy classes is the class determined by the path product of their representative elements. The preceding lemma insures that the product $\circ$ is a well-defined operation on $\pi_1(X, x_0)$. The set $\pi_1(X, x_0)$ with the $\circ$ operation is called the *fundamental group* of X at x_0, the *first homotopy group* of X at x_0, or the *Poincaré group* of X at x_0.

Theorem 4.2. *The set $\pi_1(X, x_0)$ is a group under the $\circ$ operation.*

PROOF. To show that $\pi_1(X, x_0)$ is a group, we must show that there is a loop c for which $[c]$ is an identity element, that each homotopy class $[\alpha]$ has an inverse $[\bar{\alpha}] = [\alpha]^{-1}$, and that the multiplication $\circ$ is associative. Let us prove each of these as a separate lemma.

Lemma A. *$\pi_1(X, x_0)$ has an identity element $[c]$ where c is the constant loop whose only value is x_0.*

PROOF. The constant loop c is defined by

$$c(t) = x_0, \qquad t \in I.$$

If α is a loop in X based at x_0, then

$$c * \alpha(t) = \begin{cases} x_0 & \text{if } 0 \le t \le \frac{1}{2} \\ \alpha(2t - 1) & \text{if } \frac{1}{2} \le t \le 1. \end{cases}$$

To show that $[c * \alpha] = [\alpha]$, we require a homotopy $H: I \times I \to X$ such that

$$H(\cdot, 0) = c * \alpha, \qquad H(\cdot, 1) = \alpha,$$
$$H(0, s) = H(1, s) = x_0, \qquad s \in I.$$

These requirements are filled by defining

$$H(t, s) = \begin{cases} x_0 & \text{if } 0 \le t \le (1 - s)/2 \\ \alpha\left(\dfrac{2t + s - 1}{s + 1}\right) & \text{if } (1 - s)/2 \le t \le 1. \end{cases}$$

After checking to see that H has the required properties, we will see how it was obtained. Note that

$$H(t, 0) = \left\{\begin{matrix} x_0 & \text{if } 0 \le t \le \frac{1}{2} \\ \alpha(2t - 1) & \text{if } \frac{1}{2} \le t \le 1 \end{matrix}\right\} = c * \alpha(t),$$

$$H(t, 1) = \left\{\begin{matrix} x_0 & \text{if } 0 \le t \le 0 \\ \alpha(t) & \text{if } 0 \le t \le 1 \end{matrix}\right\} = \alpha(t),$$

$$H(0, s) = x_0, \qquad H(1, s) = \alpha\left(\frac{2 + s - 1}{s + 1}\right) = \alpha(1) = x_0, \qquad s \in I.$$

Continuity of H is assured by the Continuity Lemma since $(2t + s - 1)$ divided by $(s + 1)$ is a continuous function of (t, s) and the two parts of the definition of H agree when $t = (1 - s)/2$.

The homotopy H was obtained from the diagram shown in Figure 4.3 by the analysis that follows. To define a homotopy H on the unit square which

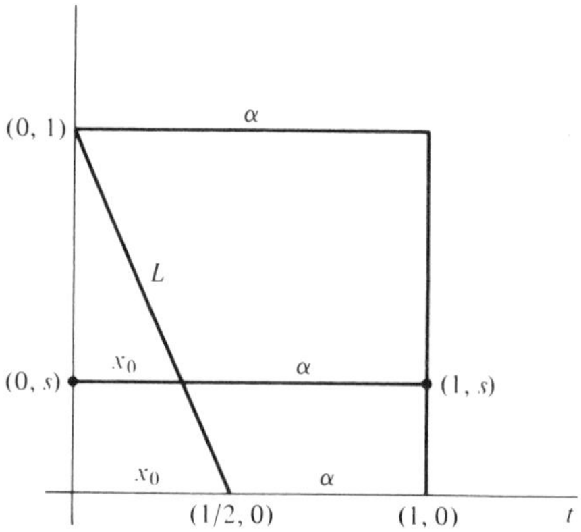

Figure 4.3

agrees with $c * \alpha$ on the bottom and with α on the top, let us intuitively assume that we will define the s-level $H(\cdot, s)$ to have value x_0 at each point (t, s) from $t = 0$ out to the diagonal line L. Then we wish $H(\cdot, s)$ to follow the route of α. Since L has equation $t = (1 - s)/2$ and the "time" remaining when $t = (1 - s)/2$ is

$$1 - \frac{(1 - s)}{2} = \frac{1 + s}{2},$$

the desired effect is accomplished by defining

$$H(t, s) = \begin{cases} x_0 & \text{if } 0 \le t \le (1 - s)/2 \\ \alpha\left(\left(t - \dfrac{(1 - s)}{2}\right) \cdot \dfrac{2}{1 + s}\right) & \text{if } (1 - s)/2 \le t \le 1. \end{cases}$$

This expression reduces to the formula for H given previously.

We have now proved the following: If $[\alpha] \in \pi_1(X, x_0)$, then

$$[c] \circ [\alpha] = [c * \alpha] = [\alpha]$$

so that $[c]$ is a left identity for $\pi_1(X, x_0)$.

In order to see that $[c]$ is a right identity as well, we need to show that $[\alpha * c] = [\alpha]$. This is accomplished by the homotopy

$$H'(t, s) = \begin{cases} \alpha\left(\dfrac{2t}{s + 1}\right) & \text{if } 0 \le t \le (s + 1)/2 \\ x_0 & \text{if } (s + 1)/2 \le t \le 1. \end{cases}$$

The intuitive picture is left to the reader.

Lemma B. *For each homotopy class* $[\alpha]$ *in* $\pi_1(X, x_0)$, *the inverse of* $[\alpha]$ *with respect to the operation* $\circ$ *and the identity element* $[c]$ *is the class* $[\bar{\alpha}]$ *where* $\bar{\alpha}(t) = \alpha(1 - t)$, $t \in I$.

PROOF. The path $\bar{\alpha}(t) = \alpha(1 - t)$ is commonly called the *reverse* of the path α. It begins at $\alpha(1) = x_0$ and traces the route of α backwards. We must prove that

$$[\alpha] \circ [\bar{\alpha}] = [\bar{\alpha}] \circ [\alpha] = [c].$$

Note that

$$[\alpha] \circ [\bar{\alpha}] = [\alpha * \bar{\alpha}],$$

$$\alpha * \bar{\alpha}(t) = \begin{cases} \alpha(2t) & \text{if } 0 \le t \le \frac{1}{2} \\ \alpha(2 - 2t) & \text{if } \frac{1}{2} \le t \le 1. \end{cases}$$

The path $\alpha * \bar{\alpha}$ follows α and then follows the reverse of α to the starting point x_0. We shall define a homotopy K for which the s-level $K(\cdot, s)$ follows route α out to $\alpha(s)$ and then retraces its steps back to x_0. This is accomplished by defining

$$K(t, s) = \begin{cases} \alpha(2ts) & \text{if } 0 \le t \le \frac{1}{2} \\ \alpha(2s - 2ts) & \text{if } \frac{1}{2} \le t \le 1. \end{cases}$$

It is easily observed that

$$K(\cdot, 0) = c, \qquad K(\cdot, 1) = \alpha * \bar{\alpha},$$
$$K(0, s) = K(1, s) = x_0, \qquad s \in I,$$

and that K is continuous.

Thus

$$[\alpha] \circ [\bar{\alpha}] = [\alpha * \bar{\alpha}] = [c],$$

so $[\bar{\alpha}]$ is a right inverse for $[\alpha]$. Since the reverse of the reverse of α is itself α (i.e., $\bar{\bar{\alpha}} = \alpha$), the same proof shows that

$$[\bar{\alpha}] \circ [\alpha] = [\bar{\alpha}] \circ [\bar{\bar{\alpha}}] = [c],$$

and hence $[\bar{\alpha}] = [\alpha]^{-1}$ is a two-sided inverse for $[\alpha]$.

Lemma C. *The multiplication $\circ$ is associative.*

Proof. Let $[\alpha]$, $[\beta]$, and $[\gamma]$ be members of $\pi_1(X, x_0)$. We must prove that

$$([\alpha] \circ [\beta]) \circ [\gamma] = [\alpha] \circ ([\beta] \circ [\gamma])$$

or, equivalently,

$$[(\alpha * \beta) * \gamma] = [\alpha * (\beta * \gamma)].$$

A little arithmetic shows that

$$(\alpha * \beta) * \gamma(t) = \begin{cases} \alpha(4t) & \text{if } 0 \le t \le \frac{1}{4} \\ \beta(4t - 1) & \text{if } \frac{1}{4} \le t \le \frac{1}{2} \\ \gamma(2t - 1) & \text{if } \frac{1}{2} \le t \le 1 \end{cases}$$

and

$$\alpha * (\beta * \gamma)(t) = \begin{cases} \alpha(2t) & \text{if } 0 \le t \le \frac{1}{2} \\ \beta(4t - 2) & \text{if } \frac{1}{2} \le t \le \frac{3}{4} \\ \gamma(4t - 3) & \text{if } \frac{3}{4} \le t \le 1. \end{cases}$$

The reader should apply the method illustrated in Lemma A to Figure 4.4, obtain the homotopy

$$L(t, s) = \begin{cases} \alpha\left(\dfrac{4t}{s + 1}\right) & \text{if } 0 \le t \le (s + 1)4 \\ \beta(4t - 1 - s) & \text{if } (s + 1)/4 \le t \le (s + 2)/4 \\ \gamma\left(\dfrac{4t - 2 - s}{2 - s}\right) & \text{if } (s + 2)/4 \le t \le 1 \end{cases}$$

and verify that it is a homotopy modulo x_0 between $(\alpha * \beta) * \gamma$ and $\alpha * (\beta * \gamma)$. This completes the proof that $\circ$ is associative and the proof of Theorem 4.2. □

The technique for obtaining the homotopies in the proof of Theorem 4.2 is extremely important in homotopy theory. The reader should be certain that

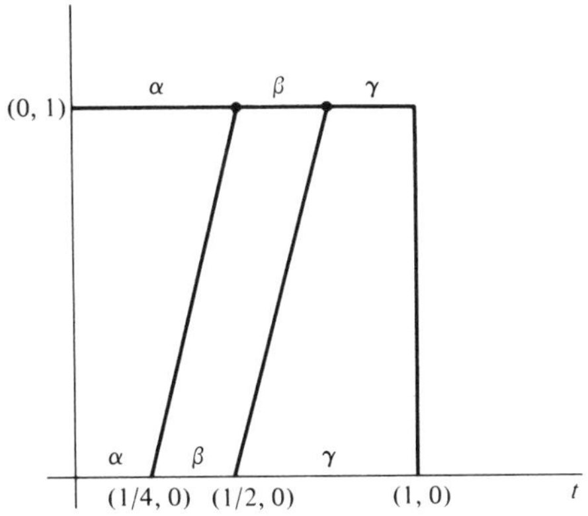

Figure 4.4

he understands the method by solving the relevant exercises at the end of the chapter.

Definition. A space X is *path connected* means that each pair of points in X can be joined by a path. In other words, if x_0 and x_1 are points in X, then there is a path in X with initial point x_0 and terminal point x_1.

Theorem 4.3. *If a space X is path connected and x_0, x_1 are points in X, then the fundamental groups $\pi_1(X, x_0)$ and $\pi_1(X, x_1)$ are isomorphic.*

Proof. Let $\rho: I \to X$ be a path such that

$$\rho(0) = x_0, \qquad \rho(1) = x_1.$$

If α is a loop based at x_0, then $(\bar{\rho} * \alpha) * \rho$ is a loop based at x_1. Here $\bar{\rho}$ denotes the reverse of ρ:

$$\bar{\rho}(t) = \rho(1 - t), \qquad 0 \leq t \leq 1.$$

We define a function $P: \pi_1(X, x_0) \to \pi_1(X, x_1)$ by

$$P([\alpha]) = [(\bar{\rho} * \alpha) * \rho], \qquad [\alpha] \in \pi_1(X, x_0).$$

It should be clear that the image of $[\alpha]$ is independent of the choice of path in $[\alpha]$ so that P is well defined.

Several observations are necessary before showing that P is an isomorphism. First, Lemma B with minor modifications shows that $[\rho * \bar{\rho}]$ and $[\bar{\rho} * \rho]$ are the identity elements of $\pi_1(X, x_0)$ and $\pi_1(X, x_1)$ respectively. Second, Lemma C can be easily modified to show that for any paths α, β, γ for which $(\alpha * \beta) * \gamma$ and $\alpha * (\beta * \gamma)$ are defined, the indicated triple products are equivalent. Thus in $[(\bar{\rho} * \alpha) * \rho]$, we may ignore the inner parentheses and simply write $[\bar{\rho} * \alpha * \rho]$ since the equivalence class is the same regardless of the way in which the terms of the product are associated.

Now consider $[\alpha]$, $[\beta]$ in $\pi_1(X, x_0)$.

$$\begin{aligned} P([\alpha] \circ [\beta]) &= P([\alpha * \beta]) = [\bar{\rho} * \alpha * \beta * \rho] = [\bar{\rho} * \alpha * \rho * \bar{\rho} * \beta * \rho] \\ &= [\bar{\rho} * \alpha * \rho] \circ [\bar{\rho} * \beta * \rho] = P([\alpha]) \circ P([\beta]). \end{aligned}$$

Thus P is a homomorphism.

The function $Q\colon \pi_1(X, x_1) \to \pi_1(X, x_0)$ defined by

$$Q([\sigma]) = [\rho * \sigma * \bar{\rho}], \qquad [\sigma] \in \pi_1(X, x_1)$$

is the inverse of P. To see this, observe that for $[\alpha] \in \pi_1(X, x_0)$,

$$\begin{aligned} QP([\alpha]) &= Q([\bar{\rho} * \alpha * \rho]) = [\rho * \bar{\rho} * \alpha * \rho * \bar{\rho}] \\ &= [\rho * \bar{\rho}] \circ [\alpha] \circ [\rho * \bar{\rho}] = [\alpha]. \end{aligned}$$

Thus QP is the identity map on $\pi_1(X, x_0)$ and, by symmetry, we observe that PQ must be the identity map on $\pi_1(X, x_1)$. Thus the indicated fundamental groups are isomorphic. □

Because of the preceding theorem, mention of a base point for the fundamental group of a path connected space is often omitted. We shall refer sometimes to "the fundamental group of X" and write $\pi_1(X)$, when X is path connected, since the same abstract group will result regardless of the choice of the base point. This applies primarily to the process of computing the fundamental group of a given space. Theorem 4.3 does not guarantee, however, that the isomorphism between $\pi_1(X, x_0)$ and $\pi_1(X, x_1)$ is unique; quite often different paths lead to different isomorphisms. For this reason, there are many applications of the fundamental group in which the specification of a base point is important. When comparing fundamental groups of two spaces X and Y on the basis of a continuous map $f\colon X \to Y$, for example, it is usually necessary to specify a base point for each space.

Definition. A path connected space X is *simply connected* provided that $\pi_1(X)$ is the trivial group.

Theorem 4.4. *Every contractible space is simply connected.*

Proof. Let X be a contractible space. There is a point x_0 in X and a homotopy $H\colon X \times I \to X$ such that

$$H(x, 0) = x, \qquad H(x, 1) = x_0, \qquad x \in X.$$

It is easy to see that X is path connected. If $x \in X$, the function

$$\alpha_x = H(x, \cdot)\colon I \to X$$

is a path from $H(x, 0) = x$ to $H(x, 1) = x_0$. Thus any two points x and y are joined by the path $\alpha_x * \bar{\alpha}_y$ where $\bar{\alpha}_y$ is the reverse of α_y.

Assume for a moment that H has the additional property

$$H(x_0, s) = x_0, \qquad s \in I.$$

For $[\alpha] \in \pi_1(X, x_0)$, define a homotopy $h\colon I \times I \to X$ by

$$h(t, s) = H(\alpha(t), s).$$

Then

$$h(t, 0) = \alpha(t), \qquad h(t, 1) = x_0, \qquad t \in I$$
$$h(0, s) = h(1, s) = x_0, \qquad s \in I.$$

Here we have used our additional assumption $H(x_0, s) = x_0$ to produce $h(0, s) = h(1, s) = x_0$. Thus h demonstrates that α is equivalent to c, the constant loop whose only value is x_0. Then $[\alpha] = [c]$ and $\pi_1(X, x_0)$ consists only of an identity element.

But what happens if the path $H(x_0, \cdot)\colon I \to X$ is not constant? We must then modify each level of the homotopy h to produce at each level a loop based at x_0. The procedure is illustrated in Figure 4.5, and the revised definition of h is left as an exercise for the reader. □

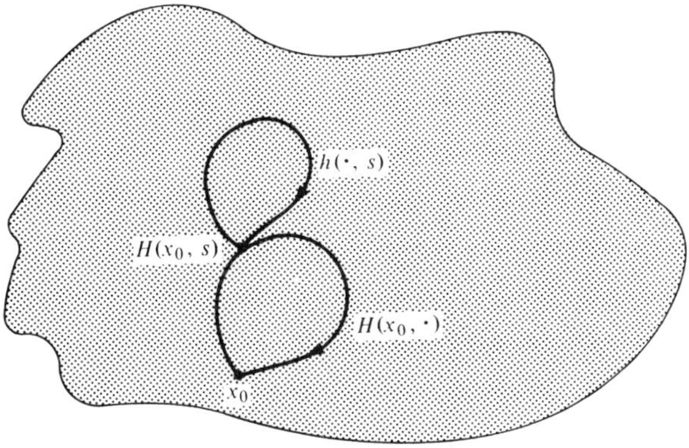

Figure 4.5

4.3 The Covering Homotopy Property for S^1

This section is devoted to determining the fundamental group of the circle. It will be convenient to consider the unit circle S^1 as a subset of the complex plane; we thus consider $\mathbb{R}^2$ as the set of all complex numbers $x = x_1 + ix_2$ where $i = \sqrt{-1}$.

We shall refer several times to the function $p\colon \mathbb{R} \to S^1$ defined by

$$p(t) = \exp(2\pi i t), \qquad t \in \mathbb{R}.$$

Here exp denotes the exponential function on the complex plane. In particular, if t is in the set $\mathbb{R}$ of real numbers, then

$$\exp(2\pi i t) = \cos(2\pi t) + i \sin(2\pi t).$$

Note that p maps each integer n in $\mathbb{R}$ to 1 in S^1 and wraps each interval $[n, n + 1]$ exactly once around S^1 in the counterclockwise direction.

Definition. If $\sigma: I \to S^1$ is a path, then a path $\tilde{\sigma}: I \to \mathbb{R}$ such that $p\tilde{\sigma} = \sigma$ is called a *covering path* of σ or a *lifting* of σ to the real line $\mathbb{R}$. If $F: I \times I \to S^1$ is a homotopy, then a homotopy $\tilde{F}: I \times I \to \mathbb{R}$ such that $p\tilde{F} = F$ is called a *covering homotopy* or a *lifting* of F.

Theorem 4.5 (The Covering Path Property). *If $\sigma: I \to S^1$ is a path in S^1 with initial point* 1, *then there is a unique covering path $\tilde{\sigma}: I \to \mathbb{R}$ with initial point* 0.

PROOF. Let U_1 denote the open arc on S^1 beginning at 1 and extending in the counterclockwise direction to $-i$, and let U_2 denote the open arc from -1 counterclockwise to i, as shown in Figure 4.6. Then U_1 and U_2 are open sets in S^1, $U_1 \cup U_2 = S^1$ and

$$p^{-1}(U_1) = \bigcup_{n=-\infty}^{\infty} (n, n + \tfrac{3}{4}),$$

$$p^{-1}(U_2) = \bigcup_{n=-\infty}^{\infty} (n - \tfrac{1}{2}, n + \tfrac{1}{4}).$$

Figure 4.6

Note that p maps each interval $(n, n + \frac{3}{4})$ homeomorphically onto U_1 and maps each interval $(n - \frac{1}{2}, n + \frac{1}{4})$ homeomorphically onto U_2.

Here is the intuitive idea behind the proof. Subdivide the range of the path σ into sections so that each section is contained either in U_1 or in U_2. If a particular section is contained in U_1, we choose one of the intervals $V = (n, n + \frac{3}{4})$ and consider the restriction $p|_V$ of p to this interval. Composing $(p|_V)^{-1}$ with this section of the path "lifts" the section to a section of a path in $\mathbb{R}$. The same method applies to sections lying in U_2. To insure continuity we must be careful that the initial point of a given lifted section be the terminal point of the lifted section that precedes it.

This method is applied inductively as follows. Let ϵ be a Lebesgue number for the open cover $\{\sigma^{-1}(U_1), \sigma^{-1}(U_2)\}$ of I. Choose a sequence

$$0 = t_0 < t_1 < t_2 < \cdots < t_n = 1$$

of numbers in I with each successive pair differing by less than ϵ. Then the image $\sigma([t_i, t_{i+1}])$ of any subinterval $[t_i, t_{i+1}]$, $0 \leq i \leq n - 1$, must be contained in either U_1 or U_2.

Now, $\sigma([t_0, t_1])$ must be contained in U_2 since

$$\sigma(t_0) = \sigma(0) = 1 \notin U_1.$$

Let $V_1 = (-\frac{1}{2}, \frac{1}{4})$ and define $\tilde{\sigma}$ on $[t_0, t_1]$ by

$$\tilde{\sigma}(t) = (p|_{V_1})^{-1}\sigma(t).$$

Proceeding inductively, suppose that σ has been defined on the interval $[t_0, t_k]$. Then

$$\sigma([t_k, t_{k+1}]) \subset U$$

where U is either U_1 or U_2. Let V_{k+1} be the component of $p^{-1}(U)$ to which $\tilde{\sigma}(t_k)$ belongs. Note that V_{k+1} is one of the intervals $(n, n + \frac{3}{4})$ or $(n - \frac{1}{2}, n + \frac{1}{4})$. Then $p|_{V_{k+1}}$ is a homeomorphism, and the desired extension of $\tilde{\sigma}$ to $[t_k, t_{k+1}]$ is obtained by defining

$$\tilde{\sigma}(t) = (p|_{V_{k+1}})^{-1}\sigma(t), \qquad t \in [t_k, t_{k+1}].$$

The continuity of $\tilde{\sigma}$ is guaranteed by the Continuity Lemma since the lifted sections agree at the endpoints t_k. This inductive step extends the definition of $\tilde{\sigma}$ to $[t_0, t_n] = I$.

To prove that $\tilde{\sigma}$ is the only such covering path, suppose that σ' also satisfies the required properties $p\sigma' = \sigma$ and $\sigma'(0) = 0$. Then the path $\tilde{\sigma} - \sigma'$ has initial point 0 and

$$p(\tilde{\sigma}(t) - \sigma'(t)) = p\tilde{\sigma}(t)/p\sigma'(t) = \sigma(t)/\sigma(t) = 1, \qquad t \in I,$$

so $\tilde{\sigma} - \sigma'$ is a covering path of the constant path whose only value is 1. Since p maps only integers to 1, then $\tilde{\sigma} - \sigma$ must have only integral values. Thus, since I is connected, $\tilde{\sigma} - \sigma'$ can have only one integral value. This one value must be the initial value, 0. Therefore $\tilde{\sigma} - \sigma' = 0$, so $\tilde{\sigma} = \sigma'$. The required lifting $\tilde{\sigma}$ is therefore unique. □

Corollary (The Generalized Covering Path Property). *If σ is a path in S^1 and r is a real number such that $p(r) = \sigma(0)$, then there is a unique covering path $\tilde{\sigma}$ of σ with initial point r.*

PROOF. The path $\sigma/\sigma(0)$ is a path in S^1 with initial point $\sigma(0)/\sigma(0) = 1$ and therefore has a unique covering path η with initial point 0. The path $\tilde{\sigma}: I \to \mathbb{R}$ defined by

$$\tilde{\sigma}(t) = r + \eta(t), \qquad t \in I,$$

is the required covering path of σ with initial point r. The uniqueness of $\tilde{\sigma}$ follows from that of η. □

Theorem 4.6 (The Covering Homotopy Property). *If $F: I \times I \to S^1$ is a homotopy such that $F(0, 0) = 1$, then there is a unique covering homotopy $\tilde{F}: I \times I \to \mathbb{R}$ such that $\tilde{F}(0, 0) = 0$.*

PROOF. The proof is similar to that of the Covering Path Property; in fact, we use the same open sets U_1, U_2 in S^1. By a Lebesgue number argument, there must exist numbers

$$0 = t_0 < t_1 < \cdots < t_n = 1, \qquad 0 = s_0 < s_1 < \cdots < s_m = 1$$

such that F maps any rectangle $[t_i, t_{i+1}] \times [s_k, s_{k+1}]$ into either U_1 or U_2. Since

$$F(0, 0) = 1 \notin U_1,$$

then $F([t_0, t_1] \times [s_0, s_1])$ must be contained in U_2. Let $V_1 = (-\frac{1}{2}, \frac{1}{4})$ and define $\tilde{F}$ on $[t_0, t_1] \times [s_0, s_1]$ by

$$F(t, s) = (p|_{V_1})^{-1}F(t, s).$$

Now extend the definition of F over the rectangles $[t_i, t_{i+1}] \times [s_0, s_1]$ in succession as in the proof of the Covering Path Property, being careful that the definitions agree on common edges of adjacent rectangles. This defines F on the strip $I \times [s_0, s_1]$.

Proceeding inductively, suppose that F has been defined on $(I \times [s_0, s_k]) \cup ([t_0, t_i] \times [s_k, s_{k+1}])$. We wish to extend the domain to include $[t_i, t_{i+1}] \times [s_k, s_{k+1}]$, as shown in Figure 4.7. Let

$$A = \{(x, y) \in [t_i, t_{i+1}] \times [s_k, s_{k+1}] : x = t_i \text{ or } y = s_k\}$$

be the common boundary of the present domain of F and $[t_i, t_{i+1}] \times [s_k, s_{k+1}]$. Now, $F([t_i, t_{i+1}] \times [s_k, s_{k+1}])$ is contained in either U_1 or U_2. Denote this containing set by U, and let V be the component of $p^{-1}(U)$ which contains $\tilde{F}(A)$. Define $\tilde{F}$ on $[t_i, t_{i+1}] \times [s_k, s_{k+1}]$ by

$$\tilde{F}(t, s) = (p|_V)^{-1}F(t, s).$$

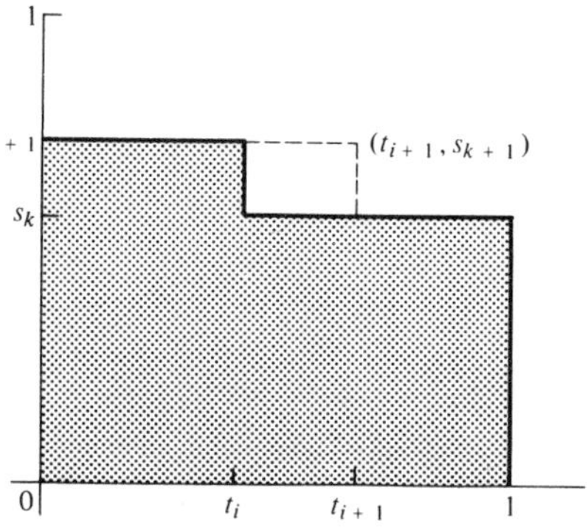

Figure 4.7

The continuity of $\tilde{F}$ follows from the Continuity Lemma since the old and new definitions of $\tilde{F}$ agree on the closed set A. This induction extends the domain of $\tilde{F}$ to $[t_0, t_n] \times [s_0, s_m] = I \times I$.

To see that $\tilde{F}$ is the only covering homotopy of F with $\tilde{F}(0, 0) = 0$, suppose that F' is another one. Then the homotopy $\tilde{F} - F'$ has the properties

$$(\tilde{F} - F')(0, 0) = \tilde{F}(0, 0) - F'(0, 0) = 0,$$
$$p(\tilde{F} - F')(t, s) = p\tilde{F}(t, s)/pF'(t, s) = F(t, s)/F(t, s) = 1,$$

for all (t, s) in $I \times I$. Thus, as in the case of covering paths, $F - F'$ can have only one integral value, namely 0. Then $F = F'$ and the covering homotopy is unique. □

Definition. Let α be a loop in S^1 with base point 1. The Covering Path Property insures that there is exactly one covering path $\tilde{\alpha}$ of α with initial point 0. Since

$$1 = \alpha(1) = p\tilde{\alpha}(1) = \exp(2\pi i\tilde{\alpha}(1)),$$

then $\tilde{\alpha}(1)$ must be an integer. This integer is called the *degree* of the loop α.

Theorem 4.7. *Two loops α and β in S^1 with base point* 1 *are equivalent if and only if they have the same degree.*

PROOF. Let $\tilde{\alpha}$ and $\tilde{\beta}$ denote the covering paths of α and β respectively having initial point 0 in $\mathbb{R}$.

Suppose first that α and β have the same degree so that $\tilde{\alpha}(1) = \tilde{\beta}(1)$. Define a homotopy $H: I \times I \to \mathbb{R}$ by

$$H(t, s) = (1 - s)\tilde{\alpha}(t) + s\tilde{\beta}(t), \qquad (t, s) \in I \times I.$$

Then H demonstrates the equivalence of $\tilde{\alpha}$ and $\tilde{\beta}$ as paths in $\mathbb{R}$. Note in particular that $H(1, s)$ is the common degree of α and β for each s in I. The homotopy

$$pH: I \times I \to S^1$$

shows the equivalence of α and β as loops in S^1.

Suppose now that α and β are equivalent loops in S^1 and that $K: I \times I \to S^1$ is a homotopy such that

$$K(\cdot, 0) = \alpha, \qquad K(\cdot, 1) = \beta,$$
$$K(0, s) = K(1, s) = 1, \qquad s \in I.$$

By the Covering Homotopy Property, there is a covering homotopy $\tilde{K}: I \times I \to \mathbb{R}$ such that

$$\tilde{K}(0, 0) = 0, \qquad p\tilde{K} = K.$$

Then

$$p\tilde{K}(0, s) = K(0, s) = 1, \qquad s \in I,$$

so $\tilde{K}(0, s)$ must be an integer for each value of s. Since I is connected, $\tilde{K}(0, \cdot)$ must have only the value $\tilde{K}(0, 0) = 0$. A similar argument shows that $\tilde{K}(1, \cdot)$ is also a constant function.

Since

$$p\tilde{K}(\cdot, 0) = K(\cdot, 0) = \alpha, \qquad p\tilde{K}(\cdot, 1) = K(\cdot, 1) = \beta,$$

then $\tilde{K}(\cdot, 0) = \tilde{\alpha}$ and $\tilde{K}(\cdot, 1) = \tilde{\beta}$ are the unique covering paths of α and β respectively with initial point 0. Thus

$$\text{degree } \alpha = \tilde{\alpha}(1) = \tilde{K}(1, 0) = \tilde{K}(1, 1) = \tilde{\beta}(1) = \text{degree } \beta,$$

so α and β must have the same degree. □

Corollary. *The fundamental group $\pi_1(S^1)$ is isomorphic to the group $\mathbb{Z}$ of integers under addition.*

PROOF. Consider $\pi_1(S^1, 1)$, and define a function

$$\deg\colon \pi_1(S^1, 1) \to \mathbb{Z}$$

by

$$\deg[\alpha] = \text{degree } \alpha.$$

The preceding theorem insures that deg is well-defined and one-to-one.

To see that deg maps $\pi_1(S^1, 1)$ onto $\mathbb{Z}$, let n be an integer. The loop γ in S^1 defined by

$$\gamma(t) = \exp(2\pi int)$$

is covered by the path

$$t \to nt, \qquad t \in I,$$

and therefore has degree n. Thus $\deg[\gamma] = n$.

Suppose now that $[\alpha]$ and $[\beta]$ are in $\pi_1(S^1, 1)$. We must show that

$$\deg([\alpha] \circ [\beta]) = \deg[\alpha] + \deg[\beta].$$

If $\tilde{\alpha}$ and $\tilde{\beta}$ are the unique covering paths of α and β which begin at 0, then the path $f\colon I \to \mathbb{R}$ defined by

$$f(t) = \begin{cases} \tilde{\alpha}(2t) & \text{if } 0 \le t \le \frac{1}{2} \\ \tilde{\alpha}(1) + \tilde{\beta}(2t - 1) & \text{if } \frac{1}{2} \le t \le 1 \end{cases}$$

is the covering path of $\alpha * \beta$ with initial point 0. Thus $\text{degree}(\alpha * \beta) = f(1) = \tilde{\alpha}(1) + \tilde{\beta}(1) = \text{degree } \alpha + \text{degree } \beta$. Then

$$\begin{aligned} \deg([\alpha] \circ [\beta]) &= \text{degree}(\alpha * \beta) = \text{degree } \alpha + \text{degree } \beta \\ &= \deg[\alpha] + \deg[\beta]. \end{aligned}$$

□

The most important topic of this section has been the Covering Homotopy Property. We shall see it again in a more general form in Chapter 5, and those who take additional courses in algebraic topology will find that it is one of the most useful concepts in homotopy theory.

4.4 Examples of Fundamental Groups

We now know that the fundamental group of a circle is the group of integers and that the fundamental group of any contractible space is trivial. The observant reader has probably surmised that the fundamental group is difficult to compute, even for simple spaces.

Homeomorphic spaces have isomorphic fundamental groups. The proof of this fact is left as an exercise. In this section we shall present less stringent conditions which insure that two spaces have isomorphic fundamental groups. This will allow us to determine the fundamental groups of several spaces similar to S^1. In the latter part of the section we shall prove a theorem which shows that the fundamental group of the n-sphere S^n is trivial for $n > 1$.

Definition. Let X be a space and A a subspace of X. Then A is a *deformation retract* of X means that there is a homotopy $H: X \times I \to X$ such that

$$H(x, 0) = x, \qquad H(x, 1) \in A, \qquad x \in X,$$
$$H(a, t) = a, \qquad a \in A,\ t \in I.$$

The homotopy H is called a *deformation retraction.*

Theorem 4.8. *If A is a deformation retract of a space X and x_0 is a point of A, then $\pi_1(X, x_0)$ is isomorphic to $\pi_1(A, x_0)$.*

Proof. Let $H: X \times I \to X$ be a deformation retraction of X onto A. Then if α is a loop in X with base point x_0, $H(\alpha(\cdot), 1)$ is a loop in A with base point x_0. We therefore define $h: \pi_1(X, x_0) \to \pi_1(A, x_0)$ by

$$h([\alpha]) = [H(\alpha(\cdot), 1)].$$

For $[\alpha]$, $[\beta]$ in $\pi_1(X, x_0)$,

$$\begin{aligned} h([\alpha] \circ [\beta]) &= h([\alpha * \beta]) = [H(\alpha * \beta(\cdot), 1)] = [H(\alpha(\cdot), 1) * H(\beta(\cdot), 1)] \\ &= h([\alpha]) \circ h([\beta]), \end{aligned}$$

so h is a homomorphism.

The fact that $H(\alpha(\cdot), 1)$ is equivalent to $H(\alpha(\cdot), 0) = \alpha$ as loops in X insures that h is one-to-one. If $[\gamma]$ is in $\pi_1(A, x_0)$, then γ determines a homotopy class (still called $[\gamma]$) in $\pi_1(X, x_0)$. Since H leaves each point of A fixed, then

$$h([\gamma]) = H(\gamma(\cdot), 1) = [\gamma],$$

so h maps $\pi_1(X, x_0)$ onto $\pi_1(A, x_0)$. This completes the proof that h is an isomorphism. □

Example 4.2. Consider the punctured plane $\mathbb{R}^2\backslash\{p\}$ consisting of all points in $\mathbb{R}^2$ except a particular point p. Let A be a circle with center p as shown in Figure 4.8.

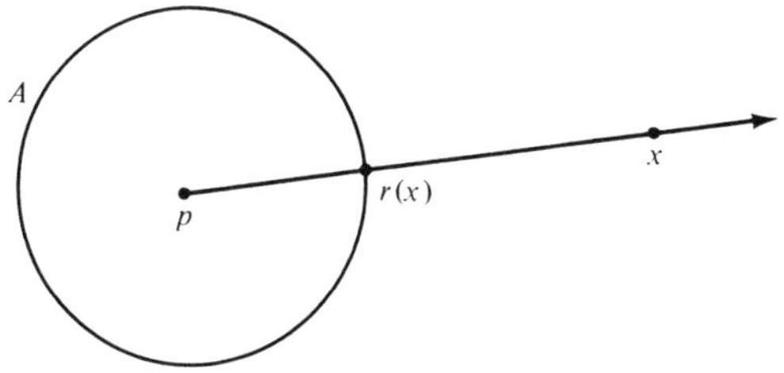

Figure 4.8

For $x \in \mathbb{R}^2\backslash\{p\}$, the half line from p through x intersects the circle A at a point $r(x)$. This function r is clearly a retraction of $\mathbb{R}^2\backslash\{p\}$ onto A. Define a homotopy $H: (\mathbb{R}^2\backslash\{p\}) \times I \to \mathbb{R}^2\backslash\{p\}$ by

$$H(x, t) = tr(x) + (1 - t)x, \qquad x \in \mathbb{R}^2\backslash\{p\},\ t \in I.$$

It is easy to see that H is a deformation retraction, so A is a deformation retract of $\mathbb{R}^2 \backslash \{p\}$. Thus

$$\pi_1(\mathbb{R}^2 \backslash \{p\}) \cong \pi_1(A) \cong \mathbb{Z}.$$

Example 4.3. Consider an annulus X in the plane. Both the inner and outer circles of X are deformation retracts, so $\pi_1(X)$ is the group of integers.

Example 4.4. Each of the following spaces is contractible, so each has fundamental group $\{0\}$:

(a) a single point,
(b) an interval on the real line,
(c) the real line,
(d) Euclidean n-space $\mathbb{R}^n$,
(e) any convex set in $\mathbb{R}^n$.

Theorem 4.9. *Let X and Y be spaces with points x_0 in X and y_0 in Y. Then*

$$\pi_1(X \times Y, (x_0, y_0)) \cong \pi_1(X, x_0) \oplus \pi_1(Y, y_0).$$

PROOF. Let p_1 and p_2 denote the projections of the product space $X \times Y$ on X and Y respectively:

$$p_1(x, y) = x, \qquad p_2(x, y) = y, \qquad (x, y) \in X \times Y.$$

Any loop α in $X \times Y$ based at (x_0, y_0) determines loops

$$\alpha_1 = p_1\alpha, \qquad \alpha_2 = p_2\alpha$$

in X and Y based at x_0 and y_0 respectively. Conversely, any pair of loops α_1 and α_2 in X and Y based at x_0 and y_0 respectively determines a loop $\alpha = (\alpha_1, \alpha_2)$ in $X \times Y$ based at (x_0, y_0). The function

$$h\colon \pi_1(X \times Y, (x_0, y_0)) \to \pi_1(X, x_0) \oplus \pi_1(Y, y_0)$$

defined by

$$h([\alpha]) = ([\alpha_1], [\alpha_2]), \qquad [\alpha] \in \pi_1(X \times Y, (x_0, y_0)),$$

is the required isomorphism. □

Example 4.5. The torus T is homeomorphic to the product $S^1 \times S^1$. Hence

$$\pi_1(T) \cong \pi_1(S^1) \oplus \pi_1(S^1) \cong \mathbb{Z} \oplus \mathbb{Z}.$$

Example 4.6. An *n-dimensional torus* T^n is the product of n unit circles. Hence $\pi_1(T^n)$ is isomorphic to the direct sum of n copies of the group of integers.

Example 4.7. A *closed cylinder* C is the product of a circle S^1 and a closed interval $[a, b]$. Thus

$$\pi_1(C) \cong \pi_1(S^1) \oplus \pi_1([a, b]) \cong \mathbb{Z} \oplus \{0\} \cong \mathbb{Z}.$$

Theorem 4.10. *Let X be a space for which there is an open cover $\{V_i\}$ of X such that*

(a) $\bigcap V_i \neq \varnothing$,
(b) *each V_i is simply connected, and*
(c) *for $i \neq j$, $V_i \cap V_j$ is path connected. Then X is simply connected.*

PROOF. Since each of the open sets V_i is path connected and their intersection is not empty, it follows easily that X is path connected. Let x_0 be a point in $\bigcap V_i$. We must show that $\pi_1(X, x_0)$ is the trivial group.

Let $[\alpha]$ be a member of $\pi_1(X, x_0)$. Then $\alpha: I \to X$ is a continuous map, so the set of all inverse images $\{\alpha^{-1}(V_i)\}$ is an open cover of the unit interval I. Since I is compact, this open cover has a Lebesgue number ϵ. Then there is a partition

$$0 = t_0 < t_1 < t_2 < \cdots < t_n = 1$$

of I such that if $0 \leq j \leq n - 1$, then $\alpha([t_j, t_{j+1}])$ is a subset of some V_i. (We need only require that successive terms of the partition differ by less than ϵ.)

Let us alter the notation of the open cover $\{V_i\}$, if necessary, so that

$$\alpha([t_j, t_{j+1}]) \subset V_j, \qquad 0 \leq j \leq n - 1.$$

Letting

$$\alpha_j(s) = \alpha((1 - s)t_j + st_{j+1}), \qquad s \in I,$$

we have a sequence $\{\alpha_j\}_{j=0}^{n-1}$ of paths in X such that $\alpha_j(I)$ is a subset of the simply connected set V_j, and

$$[\alpha] = [\alpha_0 * \alpha_1 * \alpha_2 * \cdots * \alpha_{n-1}].$$

This process is illustrated for $n = 4$ in Figure 4.9.

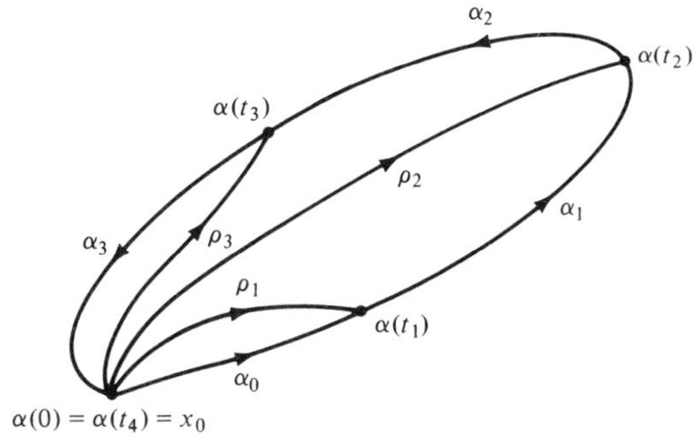

Figure 4.9

Since $V_{j-1} \cap V_j$ is path connected, there is a path ρ_j from x_0 to $\alpha(t_j)$, $1 \leq j \leq n - 1$, lying entirely in $V_{j-1} \cap V_j$. (Note that $\alpha(t_j)$ is the terminal point of α_{j-1} and the initial point of α_j.) Since the product $\bar{\rho}_j * \rho_j$ of ρ_j and its reverse is equivalent to the constant loop at x_0, then

$$\begin{aligned}[\alpha] &= [\alpha_0 * \bar{\rho}_1 * \rho_1 * \alpha_1 * \bar{\rho}_2 * \rho_2 * \alpha_2 * \cdots * \bar{\rho}_{n-1} * \rho_{n-1} * \alpha_{n-1}] \\ &= [\alpha_0 * \bar{\rho}_1] \circ [\rho_1 * \alpha_1 * \bar{\rho}_2] \circ \cdots \circ [\rho_{n-2} * \alpha_{n-2} * \bar{\rho}_{n-1}] \circ [\rho_{n-1} * \alpha_{n-1}].\end{aligned}$$

The term in this product determined by α_j is the homotopy class of a loop lying in the simply connected set V_j. Hence each term of the product represents the identity class, so $[\alpha]$ must be the identity class as well. Thus $\pi_1(X) = \{0\}$, and X is simply connected. □

Example 4.8. It is left as an exercise for the reader to show that S^n, $n > 1$, has an open cover with two members satisfying the requirements of Theorem 4.10. It then follows that $\pi_1(S^n) = \{0\}$ for $n > 1$.

4.5 The Relation between $H_1(K)$ and $\pi_1(|K|)$

The fundamental group is defined for every topological space, and we have defined homology groups for polyhedra. If $|K|$ is a polyhedron with triangulation K, how are the groups $H_1(K)$ and $\pi_1(|K|)$ related? For our examples thus far (interval, circle, torus, cylinder, annulus, and n-sphere), $\pi_1(|K|)$ and $H_1(K)$ are isomorphic. This is not true in general. The precise answer is given by Theorem 4.11 which is stated here with only an outline of the proof. Complete proofs can be found in [2], Section 8-3 and in [6], Section 12.

Theorem 4.11. *If K is a connected complex, then $H_1(K)$ is isomorphic to the quotient group $\pi_1(|K|)/F$ where F is the commutator subgroup of $\pi_1(|K|)$. Thus whenever $\pi_1(|K|)$ is abelian, $\pi_1(|K|)$ and $H_1(K)$ are isomorphic.*

Outline of Proof. Choose a vertex v of K as the base point for the fundamental group. For each oriented 1-simplex σ_i of K, let α_i denote a linear homeomorphism from $[0, 1]$ onto σ_i; the α_i are called *elementary edge paths*. An *edge loop* is a product of elementary edge paths with v as initial point and terminal point. Note that an edge loop $\alpha_1 * \alpha_2 * \cdots * \alpha_n$ corresponds in a natural way to a 1-cycle $1 \cdot \sigma_1 + 1 \cdot \sigma_2 + \cdots + 1 \cdot \sigma_n$.

Although we shall not go into the lengthy details, it is true that (a) if an edge loop is equivalent to the constant loop at v, then the corresponding 1-cycle is a boundary; (b) if two edge loops are equivalent, then their corresponding 1-cycles are homologous; and (c) each loop in $|K|$ with base point v is equivalent to an edge loop.

A homomorphism.

$$f\colon \pi_1(|K|, v) \to H_1(K)$$

may now be defined as follows: For $[\alpha] \in \pi_1(|K|, v)$, let $\hat{\alpha} = \alpha_1 * \alpha_2 * \cdots * \alpha_n$ be an edge loop equivalent to α. Define the value $f([\alpha])$ to be the homology class determined by the 1-cycle which corresponds to $\hat{\alpha}$. Then f is a homomorphism from $\pi_1(|K|, v)$ onto $H_1(K)$ whose kernel is the commutator subgroup F. It follows from the First Homomorphism Theorem (Appendix 3) that the quotient group $\pi_1(|K|, v)/F$ is isomorphic to $H_1(K)$. □

The fundamental group was defined by Poincaré in *Analysis Situs*, the same paper in which he introduced homology theory, and the relation between homology and homotopy given in Theorem 4.11 was known to him.

Poincaré did not prove the relation, but he stated in *Analysis Situs* that "fundamental equivalence" of paths in the homotopy sense corresponded precisely to homological equivalence of 1-chains except for commutativity. Since the commutator subgroup F of a group G is the smallest subgroup for which G/F is abelian, it is sometimes said that $H_1(K)$ is "$\pi_1(|K|)$ made abelian."

Both the homology and homotopy relations investigate the structure of a topological space by examining the connectivity or "holes in the space." Note that homotopy is more easily defined and conceptually simpler. It does not require elaborate explanations of chains, boundaries, cycles, or quotient groups. Homotopy applies immediately to general topological spaces and does not require the special polyhedral structure that we used for homology. Thus homotopy has some real advantages over homology.

Taking the other point of view, homology is in some ways preferable to homotopy. The fundamental group is difficult to determine rigorously, even for simple spaces. Recall, for example, our computation of $\pi_1(S^1)$ and the proof of Theorem 4.4 showing that each contractible space is simply connected. We found in Chapter 2 that homology groups are effectively calculable, for pseudomanifolds at least, because of the simplicial structure of the underlying complexes. Note also that the fundamental group overlooks the existence of higher dimensional holes in S^n, $n > 1$. To describe higher dimensional connectivity by the homotopy concept, we need a generalization of the fundamental group to higher dimensions. That is to say, we need homotopy analogues of the higher dimensional homology groups. After giving some applications of the fundamental group in Chapter 5, we shall study the higher homotopy groups in Chapter 6.

In defining the homology and homotopy relations, Poincaré hoped to give an algebraic system of topological invariants that could be used to classify topological spaces, especially manifolds. Ideally, one would hope for a sequence of groups which are reasonably amenable to computation and have the property that two spaces are homeomorphic if and only if their corresponding groups are isomorphic. As pointed out earlier (Theorem 2.11), the homology characters, and thus the homology groups, provide such a classification for 2-manifolds. Poincaré's hope that the homology groups would provide a similar classification for 3-manifolds was not fulfilled. Poincaré himself showed in 1904 that two 3-manifolds may have isomorphic homology groups and not be homeomorphic. More specifically, he found a 3-manifold whose homology groups are isomorphic to those of the 3-sphere S^3 but which is not simply connected, and therefore not homeomorphic to S^3.

Poincaré was greatly preoccupied with the classification problem. He hoped that the fundamental group would overcome the deficiencies of homology theory in the classification of 3-manifolds. It does not, however, for J. W. Alexander showed in 1919, seven years after Poincaré's death, that there exist nonhomeomorphic 3-manifolds having isomorphic homology groups and isomorphic fundamental groups [26]. Alexander's examples involved fundamental groups of order five and left unanswered the famous Poincaré Conjecture:

The Poincaré Conjecture. *Every simply connected* 3-*manifold is homeomorphic to the* 3-*sphere.*

The classification problem, even for 3-manifolds, and the Poincaré Conjecture remain unsolved. Nonetheless, the fundamental group has been a powerful tool and a great stimulus for research in algebraic topology. It seems to lie at the very base of many difficult mathematical problems. We shall see some of its power as we study an important class of spaces, the covering spaces, in Chapter 5.

EXERCISES

1. Prove the Continuity Lemma.

2. Show that multiplication in $\pi_1(X, x_0)$ is well defined. In other words, show that if $\alpha \sim_{x_0} \alpha'$ and $\beta \sim_{x_0} \beta'$, then
$$\alpha * \beta \sim_{x_0} \alpha' * \beta'.$$

3. Complete the details in the proofs of Lemmas A and C.

4. Given a space X and loops α, β, γ, and δ with base point x_0 in X, exhibit a homotopy which shows that
$$(\alpha * \beta) * (\gamma * \delta) \sim_{x_0} \alpha * ((\beta * \gamma) * \delta).$$

5. Let α and β be paths in a space X both having initial point x_0 and terminal point x_1. Prove that α is equivalent to β if and only if the product $\alpha * \bar{\beta}$ of α and the reverse of β is equivalent to the constant loop at x_0.

6. Let ρ be a loop in X with base point x_0. Prove that the induced homomorphism given by the proof of Theorem 4.3,
$$P: \pi_1(X, x_0) \to \pi_1(X, x_0),$$
is the identity isomorphism if and only if the homotopy class $[\rho]$ belongs to the center of $\pi_1(X, x_0)$.

7. Let ρ and ρ' be paths in a space X both having initial point x_0 and terminal point x_1. Give a necessary and sufficient condition that the homomorphisms induced by ρ and ρ' in the proof of Theorem 4.3 be identical. Prove that your condition is correct.

8. Complete the proof of Theorem 4.4.

9. Give an example of a simply connected space which is not contractible.

10. Give an example of a contractible space X and a point x_0 in X for which there is no contraction of X to x_0 which leaves x_0 fixed throughout the contracting homotopy.

11. In analogy with the Generalized Covering Path Property, state and prove a "Generalized Covering Homotopy Property" for S^1.

12. Prove that a path connected space is simply connected if and only if every pair of paths in X having common initial point and common terminal point are equivalent.

13. Let $f: X \to Y$ be a continuous function. Prove that the function $f_*: \pi_1(X, x_0) \to \pi_1(Y, f(x_0))$ defined by

$$f_*([\alpha]) = [f\alpha], \qquad [\alpha] \in \pi_1(X, x_0),$$

is a homomorphism. Show in particular that f_* is well-defined.

14. Prove that homeomorphic spaces have isomorphic fundamental groups.

15. In the proof of Theorem 4.5, explain why the covering path $\tilde{\alpha}$ has initial point 0.

16. Explain why the loop $\gamma_n: I \to S^1$ defined by

$$\gamma_n(t) = \exp(2\pi i n t), \qquad t \in I,$$

has degree n for each integral value of n.

17. Determine the fundamental group of the Möbius strip.

18. Prove that every deformation retract of a space X is a rectract of X. Show by example that the converse is false.

19. Let X be a space consisting of two 2-spheres joined at a point. Prove that $\pi_1(X) = \{0\}$.

20. Let X be a space consisting of two circles joined at a point. Prove that $\pi_1(X)$ is a free group on two generators and hence that there are nonabelian fundamental groups.

21. Show that the function h in the proof of Theorem 4.9 is an isomorphism.

22. Show that the n-sphere S^n, $n > 1$, satisfies the hypotheses of Theorem 4.10 and that $\pi_1(S^n) = \{0\}$.

23. Prove that each of the following spaces is contractible:
(a) the real line,
(b) a convex set in $\mathbb{R}^n$,
(c) the upper hemisphere H of S^n: $H = \{(x_1, \ldots, x_{n+1}) \in S^n : x_{n+1} \geq 0\}$,
(d) $S^n \backslash \{p\}$ where p is a particular point in S^n.

24. Let p be a point in S^1. Prove that $S^1 \times \{p\}$ is a retract but not a deformation retract of $S^1 \times S^1$.

25. Prove that the fundamental group of punctured n-space $\mathbb{R}^n \backslash \{p\}$ is trivial for $n > 2$.

26. Let G be a topological group with identity element e. If α, β are loops in G with base point e, we can define a new product $\cdot$ by

$$\alpha \cdot \beta(t) = \alpha(t)\beta(t)$$

where juxtaposition of $\alpha(t)$ and $\beta(t)$ indicates their group product in G.
(a) Prove that the operation $\cdot$ on loops based at e induces a group operation on $\pi_1(G, e)$.

(b) Show that the operation induced by $\cdot$ is exactly the same as the usual product $\circ$ on $\pi_1(G, e)$. (*Hint*: Prove that $(\alpha * c)\cdot(c * \beta) = \alpha * \beta$ where c is the constant loop at e.)

(c) Prove that $\pi_1(G, e)$ is abelian. (*Hint*: Compare $(\alpha * c)\cdot(c * \beta)$ and $(c * \alpha)\cdot(\beta * c)$.

27. If K is a complex with combinatorial components $K_1, \ldots, K_r$, how is $H_1(K)$ related to the groups $\pi_1(|K_1|), \ldots, \pi_1(|K_r|)$?

28. Give an intuitive explanation of each of the following statements:

(a) The degree of a loop α in S^1 is the number of times that α wraps the interval I around the circle.

(b) The circle has one "hole" so its fundamental group is the group $\mathbb{Z}$ of integers.

(c) The fundamental groups of a torus and a figure eight (two circles joined at a point) are not isomorphic.

29. (a) Show that a loop in a space X may be considered a continuous map from S^1 into X. (*Hint*: Consider the quotient space of I obtained by identifying 0 and 1 to a single point.)

(b) Let α be a loop in S^1. Explain the relation between the degree of α in the homotopy sense and its degree in the homology sense.

30. Let X be a space consisting of two spheres S^m and S^n, where $m, n \geq 2$, tangent at a point. Prove that $\pi_1(X) = \{0\}$.

5 Covering Spaces

This chapter is designed to show the power of the fundamental group. We shall consider a class of mappings $p: E \to B$, called "covering projections," from a "covering space" E to a "base space" B to which we can extend the Covering Homotopy Property discussed in Chapter 4. Precise definitions are given in the next section.

The fundamental group is instrumental in determining and classifying the topological spaces that can be covering spaces of a given base space B. For a large class of spaces, the possible covering spaces of B are determined by the subgroups of $\pi_1(B)$. In addition, the theory of covering spaces will allow us to determine the fundamental groups of several rather complicated spaces.

5.1 The Definition and Some Examples

Recall from Chapter 4 that a space X is *path connected* provided that each pair of points in X can be joined by a path in X. A space that satisfies this property locally is called "locally path connected."

Definition. A topological space X is *locally path connected* means that X has a basis of path connected open sets. In other words, if $x \in X$ and O is an open set containing x, then there exists an open set U containing x and contained in O such that U is path connected.

Definition. A maximal path connected subset of a space X is called a *path component*. Thus A is a path component of X means that A is path connected and is not a proper subset of any path connected subset of X. The *path components of a subset* B of X are the path components of B in its subspace topology.

It is assumed throughout this chapter that all spaces considered are path connected and locally path connected unless stated otherwise.

Definition. Let E and B be spaces and $p\colon E \to B$ a continuous map. Then the pair (E, p) is a *covering space* of B means that for each point x in B there is a path connected open set $U \subset B$ such that $x \in U$ and p maps each path component of $p^{-1}(U)$ homeomorphically onto U. Such an open set U is called an *admissible neighborhood* or an *elementary neighborhood.* The space B is the *base space* and p is a *covering projection.*

In cases where the covering projection is clearly understood, one sometimes refers to E as the covering space. We shall, however, try to avoid ambiguity by referring to the covering space properly as (E, p).

Example 5.1. Consider the map $p\colon \mathbb{R} \to S^1$ from the real line to the unit circle defined in Chapter 4:

$$p(t) = e^{2\pi it} = \cos(2\pi t) + i\sin(2\pi t), \qquad t \in \mathbb{R}.$$

Then p is a covering projection. Any proper open interval or arc on S^1 can serve as an elementary neighborhood. For the particular point 1 in S^1, let U denote the right hand open interval on S^1 from $-i$ to i. Then

$$p^{-1}(U) = \bigcup_{n=-\infty}^{\infty} (n - \tfrac{1}{4}, n + \tfrac{1}{4}),$$

and the path components of $p^{-1}(U)$ are the real intervals $(n - \frac{1}{4}, n + \frac{1}{4})$. Note that p maps each of these homeomorphically onto U, as illustrated in Figure 5.1.

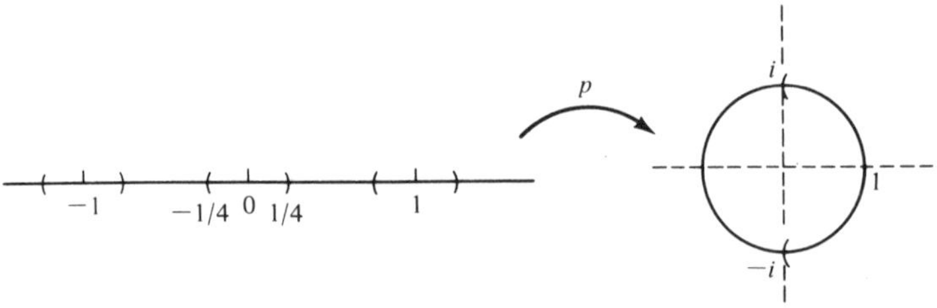

Figure 5.1

Example 5.2. For any positive integer n, let $q_n\colon S^1 \to S^1$ be the map defined by

$$q_n(z) = z^n, \qquad z \in S^1,$$

where z^n is the nth power of the complex number z. Then (S^1, q_n) is a covering space of S^1. Representing the circle in polar coordinates, the action of q_n is described as follows: q_n takes any point $(1, \theta)$ to $(1, n\theta)$. Let U be an open interval on S^1 subtended by an angle θ, $0 \leq \theta \leq 2\pi$, and containing a point x. Then $p^{-1}(U)$ consists of n open intervals each determining an angle θ/n and each containing one nth root of x. These n intervals are the path components of $p^{-1}(U)$, and each is mapped by p homeomorphically onto U. Thus any proper open interval in S^1 is an admissible neighborhood.

A repetition of Example 5.2 for negative values of n is left as an exercise.

Example 5.3. If X is a space (which, according to our assumption, must be path connected and locally path connected), then the identity map $i: X \to X$ is a covering projection, so (X, i) is a covering space of X.

Example 5.4. Let P denote the projective plane, and let $p: S^2 \to P$ be the natural map which identifies each pair of antipodal or diametrically opposite points, as in Exercise 26 of Chapter 2. To show the existence of admissible neighborhoods, let w be a point in P which is the image of two antipodal points x and $-x$. Let O be a path connected open set in S^2 containing x such that O does not contain any pair of antipodal points. (A small disc centered at x will do nicely.) Then $p(O)$ is an open set containing w, and $p^{-1}p(O)$ has path components O and the set of points antipodal to points in O. Note that p maps each of these path components homeomorphically onto $p(O)$, so $p(O)$ is an admissible neighborhood. Thus (S^2, p) is a covering space of P.

Example 5.5. Consider the map $r: \mathbb{R}^2 \to S^1 \times S^1$ from the plane to the torus defined by

$$r(t_1, t_2) = (e^{2\pi l t_1}, e^{2\pi i t_2}), \qquad (t_1, t_2) \in \mathbb{R}^2.$$

Then $(\mathbb{R}^2, r)$ is a covering space of $S^1 \times S^1$. This example is essentially a generalization of the covering projection $p: \mathbb{R} \to S^1$ of Example 1. For any point (z_1, z_2) in $S^1 \times S^1$, let U denote a small open rectangle formed by the product of two proper open intervals in S^1 containing z_1 and z_2 respectively. Then U is an admissible neighborhood whose inverse image consists of a countably infinite family of open rectangles in the plane.

Example 5.6. Let Q denote an infinite spiral, and let $q: Q \to S^1$ denote the projection described pictorially in Figure 5.2. Each point on the spiral is projected to the point on the circle directly beneath it.

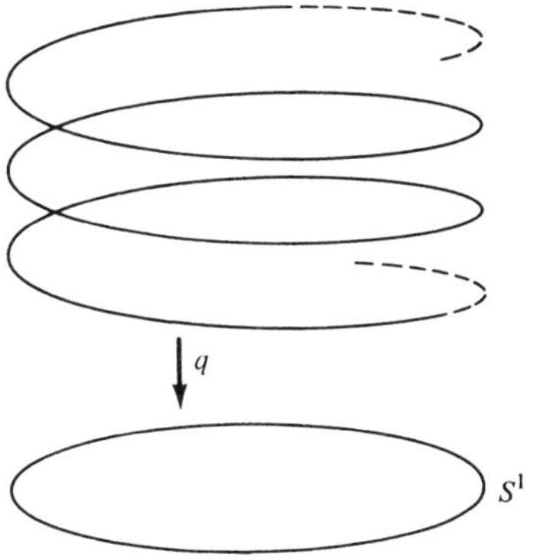

Figure 5.2

It is easy to see that (Q, q) is a covering space of S^1. In this example it is important that the spiral be infinite; a finite spiral projected in the same manner is not a covering space. By examining Figure 5.3, one can see that the points $p(x_0)$ and $p(x_1)$ lying under the ends of the spiral do not have admissible neighborhoods.

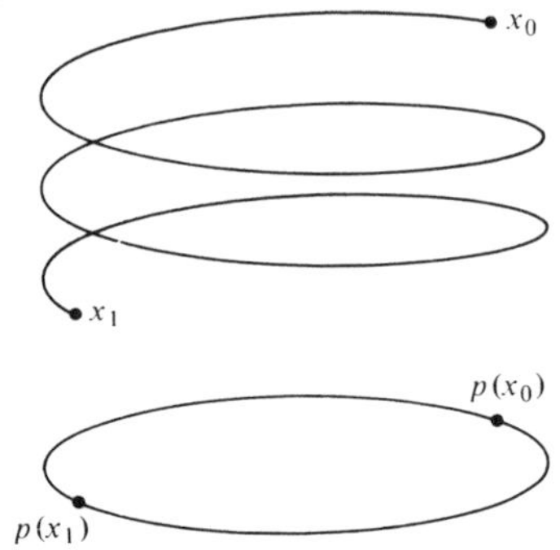

Figure 5.3

Example 5.7. The following is *not* an example of a covering space. Let R be a rectangle which is mapped by the projection onto the first coordinate to an interval A, as shown in Figure 5.4. If U is an open interval in A, then $p^{-1}(U)$ is a strip in R consisting of all points above U. This strip cannot be mapped homeomorphically onto U, so this situation does not allow admissible neighborhoods.

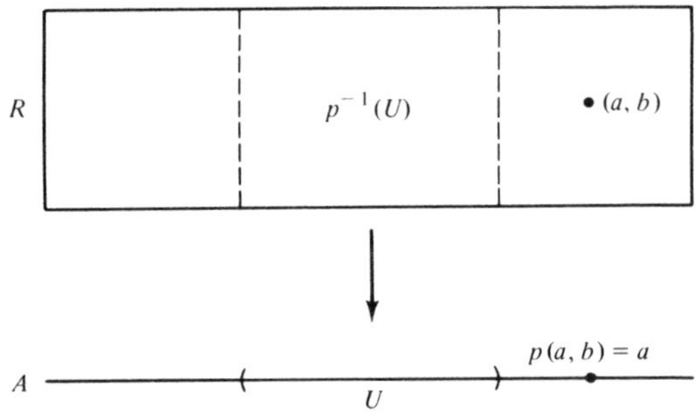

Figure 5.4

5.2 Basic Properties of Covering Spaces

In this section we shall prove some basic properties of covering spaces from the definition. The most important of these is the Covering Homotopy Property.

The following characterization of local path connectedness is left as an exercise:

Lemma. *A space X is locally path connected if and only if each path component of each open subset of X is open.*

Theorem 5.1. *Every covering projection is an open mapping.*

Proof. Let $p: E \to B$ be a covering projection. We must show that for each open set V in E, $p(V)$ is open in B. Let $x \in p(V)$, let $\tilde{x}$ be a point of V such that $p(\tilde{x}) = x$, and let U be an admissible neighborhood for x. Let W be the path component of $p^{-1}(U)$ which contains $\tilde{x}$. Since E is locally path connected, the preceding lemma implies that W is open in E. Since p maps W homeomorphically onto U, then p maps the open set $W \cap V$ to an open subset $p(W \cap V)$ in B. Thus $x \in p(W \cap V)$ and $p(W \cap V)$ is an open set contained in $p(V)$. Since x was an arbitrary point of $p(V)$, it follows that $p(V)$ is a union of open sets and is, therefore, an open set. Thus p is an open mapping. □

Theorem 5.2. *Let (E, p) be a covering space of B and X a space. If f and g are continuous maps from X into E for which $pf = pg$, then the set of points at which f and g agree is an open and closed subset of X. (We do not assume in this theorem that X is path connected or locally path connected.)*

Proof. Let $A = \{x \in X: f(x) = g(x)\}$ be the set of points at which f and g agree. To see that A is open, let x be a member of A and U an admissible neighborhood of $pf(x)$. The path component V of $p^{-1}(U)$ to which $f(x)$ belongs is an open set in E, and hence $f^{-1}(V)$ and $g^{-1}(V)$ are open in X. Since $f(x) \in V$ and $f(x) = g(x)$, then x belongs to $f^{-1}(V) \cap g^{-1}(V)$. We shall show that $f^{-1}(V) \cap g^{-1}(V)$ is a subset of A and conclude that A is open since it contains a neighborhood of each of its points.

Let $t \in f^{-1}(V) \cap g^{-1}(V)$. Then $f(t)$ and $g(t)$ are in V and are mapped by p to the common point $pf(t) = pg(t)$. Since p maps V homeomorphically onto U, it must be true that $f(t) = g(t)$. Then $t \in A$, and it follows that A is an open set.

Suppose that A is not closed, and let y be a limit point of A not in A. Then $f(y) \neq g(y)$. The point $pf(y) = pg(y)$ has an elementary neighborhood W, and $f(y)$ and $g(y)$ must belong to distinct path components V_0 and V_1 of $p^{-1}(W)$. (Why?) Since y belongs to the open set $f^{-1}(V_0) \cap g^{-1}(V_1)$, then $f^{-1}(V_0) \cap g^{-1}(V_1)$ must contain a point $t \in A$. But this is a contradiction since the point $f(t) = g(t)$ would have to belong to the disjoint sets V_0 and V_1. Thus A contains all its limit points and is a closed set. □

Corollary. *Let (E, p) be a covering space of B, and let f, g be continuous maps from a connected space X into E such that $pf = pg$. If f and g agree at a point of X, then $f = g$.*

Proof. In a connected space X, the only sets that are both open and closed are X and the empty set $\varnothing$. Thus $A = X$ or $A = \varnothing$, so f and g must be precisely equal or must disagree at every point. Note that the corollary requires only that X be connected, not path connected or locally path connected. □

Here is a situation that arises often in mathematics, particularly in topology. Suppose that spaces E and B are to be compared using a continuous map

$p: E \to B$ and that there is given another map $f: C \to B$ from a space C into B. Then a map $\tilde{f}: C \to E$ for which the diagram below is commutative, that is for which $p\tilde{f} = f$, is called a *lifting* or *covering* of f.

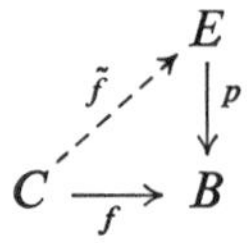

In this section we shall be interested in lifting two kinds of maps: paths and homotopies between paths. Theorem 5.2 and its corollary will be useful in showing the uniqueness of liftings.

Definition. Let (E, p) be a covering space of B, and let $\alpha: I \to B$ be a path. A path $\tilde{\alpha}: I \to E$ such that $p\tilde{\alpha} = \alpha$ is called a *lifting* or *covering path* of α. If $F: I \times I \to B$ is a homotopy, then a homotopy $\tilde{F}: I \times I \to E$ for which $p\tilde{F} = F$ is called a *lifting* or *covering homotopy* of F.

We are now ready to extend the Covering Path Property and Covering Homotopy Property that were proved earlier for the circle to covering spaces. The proofs of these important properties are merely generalizations of the proofs used in Chapter 4.

Theorem 5.3 (The Covering Path Property). *Let (E, p) be a covering space of B and $\alpha: I \to B$ a path in B beginning at a point b_0. If e_0 is a point in E with $p(e_0) = b_0$, then there is a unique covering path of α beginning at e_0.*

PROOF. Here is the basic idea of the proof: Subdivide the range of the path α into sections so that each section lies in an admissible neighborhood. If U is one of these admissible neighborhoods, then p maps each path component of $p^{-1}(U)$ homeomorphically onto U. We can then choose a path component V of $p^{-1}(U)$ and consider the restriction $p|_V$ of p to V, a homeomorphism from V onto U. Composing with $(p|_V)^{-1}$ "lifts" one section of α to E.

This method is applied inductively. Let $\{U_j\}$ be an open cover of B by admissible neighborhoods, and let ϵ be a Lebesgue number for the corresponding open cover $\{\alpha^{-1}(U_j)\}$ of I. Choose a sequence

$$0 = t_0 < t_1 < \cdots < t_n = 1$$

of numbers in I with each successive pair differing by less than ϵ. Then each subinterval $[t_i, t_{i+1}]$, $0 \le i \le n - 1$, is mapped by α into an admissible neighborhood U_{i+1}.

First consider $\alpha([t_0, t_1])$, which is contained in U_1. Let V_1 denote the path component of $p^{-1}(U_1)$ to which the desired initial point e_0 belongs. Then, for $t \in [t_0, t_1]$, define

$$\tilde{\alpha}(t) = (p|_{V_1})^{-1}\alpha(t).$$

Proceeding inductively, suppose that $\tilde{\alpha}$ has been defined on the interval $[t_0, t_k]$. Then

$$\alpha([t_k, t_{k+1}]) \subset U_{k+1},$$

so we let V_{k+1} be the path component of $p^{-1}(U_{k+1})$ to which $\tilde{\alpha}(t_k)$ belongs. Since $p|_{V_{k+1}}$ is a homeomorphism, the desired extension of $\tilde{\alpha}$ to $[t_k, t_{k+1}]$ is obtained by defining

$$\tilde{\alpha}(t) = (p|_{V_{k+1}})^{-1}\alpha(t), \qquad t \in [t_k, t_{k+1}].$$

The continuity of $\tilde{\alpha}$ follows from the Continuity Lemma since the lifted sections match properly at the end points.

The uniqueness of the covering path $\tilde{\alpha}$ can be proved from the uniqueness of each lifted section. However, it is simpler to apply the Corollary to Theorem 5.2. If α' is another covering path of α with $\alpha'(0) = e_0$, then $\tilde{\alpha}$ and α' agree at 0 and hence must be identical. □

Theorem 5.4 (The Covering Homotopy Property). *Let (E, p) be a covering space of B and $F: I \times I \to B$ a homotopy such that $F(0, 0) = b_0$. If e_0 is a point of E with $p(e_0) = b_0$, then there is a unique covering homotopy $\tilde{F}: I \times I \to E$ such that $\tilde{F}(0, 0) = e_0$.*

Having seen this property proved for a special case in Chapter 4, and having seen the proof of the Covering Path Property for covering spaces, the reader should be able to prove Theorem 5.4 for himself. A proof can be modeled after the proof of Theorem 5.3 by subdividing $I \times I$ into rectangles in the way that I was subdivided into intervals.

The Covering Homotopy Property has many important applications. One of the most important is the following criterion for determining when two paths in a covering space are equivalent.

Theorem 5.5 (The Monodromy Theorem). *Let (E, p) be a covering space of B, and suppose that $\tilde{\alpha}$ and $\tilde{\beta}$ are paths in E with common initial point e_0. Then $\tilde{\alpha}$ and $\tilde{\beta}$ are equivalent if and only if $p\tilde{\alpha}$ and $p\tilde{\beta}$ are equivalent paths in B. In particular, if $p\tilde{\alpha}$ and $p\tilde{\beta}$ are equivalent, then $\tilde{\alpha}$ and $\tilde{\beta}$ must have common terminal point.*

PROOF. If $\tilde{\alpha}$ and $\tilde{\beta}$ are equivalent by a homotopy G then the homotopy pG demonstrates the equivalence of $p\tilde{\alpha}$ and $p\tilde{\beta}$.

For a proof of the other half of the theorem, let b_0 and b_1 denote the common initial point and common terminal point respectively of $p\tilde{\alpha}$ and $p\tilde{\beta}$. Let $H: I \times I \to B$ be a homotopy demonstrating the equivalence of $p\tilde{\alpha}$ and $p\tilde{\beta}$:

$$H(\cdot, 0) = p\tilde{\alpha}, \qquad H(\cdot, 1) = p\tilde{\beta},$$
$$H(0, t) = b_0, \qquad H(1, t) = b_1, \qquad t \in I.$$

By the Covering Homotopy Property, there is a covering homotopy $\tilde{H}$ of H with $\tilde{H}(0, 0) = e_0$. Both $\tilde{\alpha}$ and the initial level $\tilde{H}(\cdot, 0)$ are covering paths of $p\tilde{\alpha}$, and they have common value e_0 at 0. Thus $\tilde{H}(\cdot, 0) = \tilde{\alpha}$ by the Corollary to Theorem 5.2. Similarly, we conclude that $\tilde{H}(\cdot, 1) = \tilde{\beta}$.

It remains to be seen that $\tilde{H}(0, \cdot)$ and $\tilde{H}(1, \cdot)$ are constant paths. But

$\tilde{H}(0, \cdot)$ is a lifting of the constant path $H(0, \cdot)$ with $\tilde{H}(0, 0) = e_0$. Since the unique lifting of a constant path is obviously a constant path, then $\tilde{H}(0, \cdot)$ must be the constant path whose only value is e_0. The same argument shows that $\tilde{H}(1, \cdot)$ must be the constant path whose only value is

$$\tilde{\alpha}(1) = \tilde{H}(1, 0) = \tilde{H}(1, 1) = \tilde{\beta}(1).$$

Thus $\tilde{H}$ is a homotopy that demonstrates the equivalence of $\tilde{\alpha}$ and $\tilde{\beta}$. □

Theorem 5.6. *If (E, p) is a covering space of B, then all the sets $p^{-1}(b)$, $b \in B$, have the same cardinal number.*

PROOF. Let b_0 and b_1 be points in B. We must define a one-to-one correspondence between $p^{-1}(b_0)$ and $p^{-1}(b_1)$. This is accomplished as follows: Let α be a path in B from b_0 to b_1. For $x \in p^{-1}(b_0)$, let $\tilde{\alpha}_x$ denote the unique covering path of α beginning at x. Then the terminal point $\tilde{\alpha}_x(1)$ is a point in $p^{-1}(b_1)$. This associates with each x in $p^{-1}(b_0)$ a point

$$f(x) = \tilde{\alpha}_x(1)$$

in $p^{-1}(b_1)$. By considering the reverse path from b_1 to b_0, one can define in the same manner a function

$$g: p^{-1}(b_1) \to p^{-1}(b_0).$$

The functions f and g are easily shown to be inverses of each other, so $p^{-1}(b_0)$ and $p^{-1}(b_1)$ must have the same cardinal number. □

Definition. If (E, p) is a covering space of B, the common cardinal number of the sets $p^{-1}(b)$, $b \in B$, is called the *number of sheets* of the covering. A covering of n sheets is called an *n-fold covering*.

Consider, for example, the covering projection $p: S^2 \to P$ of Example 5.4. Since p identifies pairs of antipodal points, the number of sheets of this covering is two. Thus (S^2, p) is referred to as the "double covering" of the projective plane.

The covering projection $p: \mathbb{R} \to S^1$ of Example 5.1 maps each integer and only the integers to $1 \in S^1$. Thus the number of sheets of this covering is countably infinite.

We close this section with a result relating the fundamental groups of E and B where (E, p) is a covering space of B. Choose base points e_0 in E and $b_0 = p(e_0)$ in B. Then if α is a loop in E based at e_0, the composition $p\alpha$ is a loop in B with base point b_0. Thus p induces a function

$$p_*: \pi_1(E, e_0) \to \pi_1(B, b_0)$$

defined by

$$p_*([\alpha]) = [p\alpha], \qquad [\alpha] \in \pi_1(E, e_0).$$

This function p_* is a group homomorphism and is called the *homomorphism induced by p*.

Theorem 5.7. *If (E, p) is a covering space of B, then the induced homomorphism $p_*: \pi_1(E, e_0) \to \pi_1(B, b_0)$ is one-to-one.*

The proof, an easy application of the Monodromy Theorem (Theorem 5.5), is left as an exercise.

5.3 Classification of Covering Spaces

The fundamental group of the base space B provides a criterion for determining when two covering spaces of B are equivalent. Each covering space determines a collection of subgroups, a conjugacy class of subgroups, of $\pi_1(B)$. We shall see that two covering spaces are homeomorphic if and only if they determine the same collection of subgroups.

Here is the terminology used in comparing covering spaces:

Definition. Let (E_1, p_1) and (E_2, p_2) be covering spaces of the same space B. A *homomorphism* from (E_1, p_1) to (E_2, p_2) is a continuous map $h: E_1 \to E_2$ for which $p_2 h = p_1$. In other words, this diagram must be commutative for h to be a homomorphism.

$$\begin{array}{ccc} E_1 & \xrightarrow{h} & E_2 \\ & {\scriptstyle p_1}\searrow \quad \swarrow{\scriptstyle p_2} & \\ & B & \end{array}$$

A homomorphism $h: E_1 \to E_2$ of covering spaces which is also a homeomorphism is called an *isomorphism*. If there is an isomorphism from one covering space to another, the two covering spaces are called *isomorphic*.

It is left as an exercise for the reader to prove that a homomorphism of covering spaces is actually a covering projection; i.e., if $h: E_1 \to E_2$ is a homomorphism, then (E_1, h) is a covering space of E_2.

Theorem 5.8. *Let (E, p) be a covering space of B. If $b_0 \in B$, then the groups $p_*\pi_1(E, e)$, as e varies over $p^{-1}(b_0)$, form a conjugacy class of subgroups of $\pi_1(B, b_0)$.*

Proof. Recall that subgroups H and K of a group G are conjugate subgroups if and only if

$$H = x^{-1}Kx$$

for some $x \in G$. The theorem then makes two assertions: (a) for any e_0, e_1 in $p^{-1}(b_0)$, the subgroups $p_*\pi_1(E, e_0)$ and $p_*\pi_1(E, e_1)$ are conjugate, and (b) any subgroup of $\pi_1(B, b_0)$ conjugate to $p_*\pi_1(E, e_0)$ must equal $p_*\pi_1(E, e)$ for some e in $p^{-1}(b_0)$.

To prove (a), consider two points e_0 and e_1 in $p^{-1}(b_0)$. Let $\rho: I \to E$ be a path from e_0 to e_1. According to Theorem 4.3, the function $P: \pi_1(E, e_0) \to \pi_1(E, e_1)$ defined by

$$P([\alpha]) = [\bar{\rho} * \alpha * \rho], \qquad [\alpha] \in \pi_1(E, e_0),$$

is an isomorphism. In particular,

$$\pi_1(E, e_1) = P\pi_1(E, e_0),$$

so

$$p_*\pi_1(E, e_1) = p_*P\pi_1(E, e_0).$$

It follows from the definition of P, however, that

$$p_*P\pi_1(E, e_0) = [p\rho]^{-1} \circ \pi_1(E, e_0) \circ [p\rho],$$

so $p_*\pi_1(E, e_0)$ and $p_*\pi_1(E, e_1)$ are conjugate subgroups of $\pi_1(B, b_0)$. Note that we are using the fact that $[p\rho]$ is an element of $\pi_1(B, b_0)$.

To prove (b), suppose that H is a subgroup conjugate to $p_*\pi_1(E, e_0)$ by some element $[\delta]$ in $\pi_1(B, b_0)$:

$$H = [\delta]^{-1} \circ p_*\pi_1(E, e_0) \circ [\delta].$$

Let $\tilde{\delta}$ be the unique covering path of δ beginning at e_0. Then $\tilde{\delta}$ has a terminal point $e \in p^{-1}(b_0)$, and the argument for part (a) shows that

$$p_*\pi_1(E, e) = [p\tilde{\delta}]^{-1} \circ p_*\pi_1(E, e_0) \circ [p\tilde{\delta}] = [\delta]^{-1} \circ p_*\pi_1(E, e_0) \circ [\delta] = H.$$

Thus

$$p_*\pi_1(E, e) = H,$$

and the set $\{p_*\pi_1(E, e): e \in p^{-1}(b_0)\}$ is precisely a conjugacy class of subgroups of $\pi_1(B, b_0)$. □

Definition. The conjugacy class of subgroups $\{p_*\pi_1(E, e): e \in p^{-1}(b_0)\}$ described in the preceding theorem is called the *conjugacy class determined by the covering space* (E, p).

The main result of this section comes next. Two covering spaces of a space B are isomorphic if and only if they determine the same conjugacy class of the fundamental group of B. We must specify a base point b_0 in B to make the representation $\pi_1(B) = \pi_1(B, b_0)$ concrete. However, according to Theorem 4.3, the choice of base point does not affect the structure of the fundamental group.

Theorem 5.9. *Let B be a space with base point b_0. Covering spaces (E_1, p_1) and (E_2, p_2) of B are isomorphic if and only if they determine the same conjugacy class of subgroups of $\pi_1(B, b_0)$.*

PROOF. The "only if" part of the proof is left as an exercise. For the "if" part, assume that the conjugacy classes of the two covering spaces are identical. Then there must be points $e_1 \in p_1^{-1}(b_0)$ and $e_2 \in p_2^{-1}(b_0)$ such that

$$p_{1*}\pi_1(E_1, e_1) = p_{2*}\pi_1(E_2, e_2).$$

The covering space isomorphism $h: E_1 \to E_2$ is defined by the following scheme: For $x \in E_1$, let α be a path in E_1 from e_1 to x. Then $p_1\alpha$ is a path in

B from b_0 to $p_1(x)$. This path has a unique covering path $\widetilde{p_1\alpha}$ in E_2 beginning at e_2 and ending at some point y in E_2. We then define $h(x) = y$. This definition is illustrated in Figure 5.5.

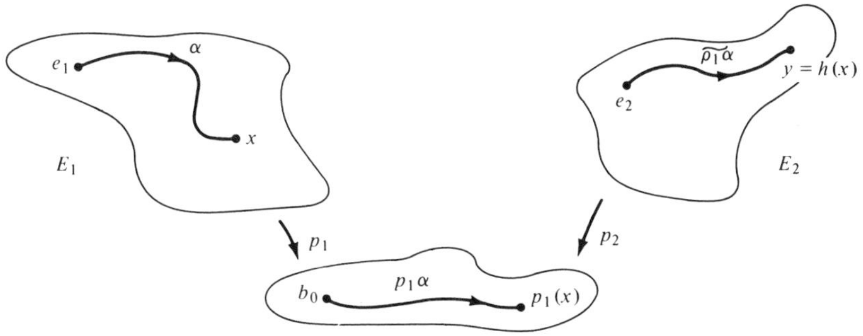

Figure 5.5

Can this h possibly be well-defined in view of the myriad choices for the path α? Does it have any chance of being continuous? The answer to both questions is "yes"; the function h is, in fact, a homeomorphism.

To show that h is well-defined, let β be another path in E_1 from e_1 to x. Since α and β both begin at e_1 and terminate at x, the product path $\alpha * \bar{\beta}$ is a loop in E_1 based at e_1. Thus

$$p_{1*}([\alpha * \bar{\beta}]) = [p_1\alpha * p_1\bar{\beta}] \in p_{1*}\pi_1(E_1, e_1).$$

But $p_{1*}\pi_1(E_1, e_1)$ and $p_{2*}\pi_1(E_2, e_2)$ are equal, so there is a member $[\gamma] \in \pi_1(E_2, e_2)$ such that

$$[p_1\alpha * p_1\bar{\beta}] = [p_2\gamma].$$

Thus the loops $p_1\alpha * p_1\bar{\beta}$ and $p_2\gamma$ are equivalent loops in B. Using the Covering Homotopy Property (Theorem 5.4) to lift a homotopy between $p_1\alpha * p_1\bar{\beta}$ and $p_2\gamma$ to E_2, we obtain a loop γ' in E_2 based at e_2 for which

$$p_2\gamma' = p_1\alpha * p_1\bar{\beta}.$$

Divide γ' into the product of two paths α' and $\bar{\beta}'$ as follows:

$$\alpha'(t) = \gamma'(t/2), \qquad \beta'(t) = \gamma'((2 - t)/2), \qquad t \in I.$$

It is a simple matter to observe that

$$p_2\alpha' = p_1\alpha, \qquad p_2\beta' = p_1\beta.$$

Since α' and β' have initial point e_2, they are the unique covering paths of $p_1\alpha$ and $p_1\beta$ with respect to the covering (E_2, p_2); i.e.,

$$\alpha' = \widetilde{p_1\alpha}, \qquad \beta' = \widetilde{p_1\beta}.$$

Then

$$\widetilde{p_1\alpha}(1) = \alpha'(1) = \gamma'(\tfrac{1}{2}), \qquad \widetilde{p_1\beta}(1) = \beta'(1) = \gamma'(\tfrac{1}{2}),$$

so the same value $h(x) = \gamma'(\frac{1}{2})$ results regardless of the choice of the path from e_1 to x. This concludes the proof that h is well-defined.

In showing that h is continuous we shall use the fact that the admissible neighborhoods form a basis for the topology of B. The proof of this is left as an exercise.

Let O be an open set in E_2 and x a member of $h^{-1}(O)$. It must be shown that there is an open set V in E_1 for which $x \in V$ and $h(V) \subset O$. Since the definition of h requires that $p_2 h = p_1$ and since p_2 is an open mapping (Theorem 5.1), then $p_1(x)$ belongs to the open set $p_2(O)$ in B. Since the admissible neighborhoods form a basis for B, there is an admissible neighborhood U such that

$$p_1(x) \in U, \qquad U \subset p_2(O).$$

Let W be the path component of $p_2^{-1}(U)$ to which $h(x)$ belongs. Then $h(x)$ belongs to the open set $O' = O \cap W$, and the restriction

$$f = p_2|O' : O' \to p_2(O')$$

is a homeomorphism. Since $p_2(O')$ is open in B, then $p_1^{-1}p_2(O')$ is open in E_1. Let V be a path connected open set in E_1 which contains x and is contained in $p_1^{-1}p_2(O')$.

To see that $h(V) \subset O$, let $t \in V$. Let α be a path in E_1 from e_1 to x and β a path in V from x to t. Then

$$h(x) = \widetilde{p_1\alpha}(1), \qquad h(t) = \widetilde{p_1\alpha * p_1\beta}(1).$$

But since $f = p_2|O'$ is a homeomorphism, the covering path of $p_1\alpha * p_1\beta$ is $\widetilde{p_1\alpha} * f^{-1}p_1\beta$. Thus

$$h(t) = f^{-1}p_1\beta(1) = f^{-1}p_1(t).$$

This point is in O' because $p_1(t) \in p_2(O')$ and f is a homeomorphism between O' and $p_2(O')$. Since $O' \subset O$, it follows that $h(t) \in O$ and hence that $h(V) \subset O$.

The proof thus far has shown that there is a covering space homomorphism h from E_1 to E_2. By looking at constant paths, it is easy to see that $h(e_1) = e_2$. The reader may be tiring at this point, especially in view of the fact that the existence of a continuous inverse for h must be shown. However, the proof thus far has essentially done that. Reversing the roles of E_1 and E_2, there must exist a continuous map $g: E_2 \to E_1$ such that

$$p_1 g = p_2, \qquad g(e_2) = e_1.$$

Consider the composite map gh from E_1 to E_1:

$$p_1 gh = p_2 h = p_1 i_1,$$

where i_1 is the identity map on E_1. Since gh and i_1 agree at e_1, the Corollary to Theorem 5.2 implies that gh is the identity map on E_1. By symmetry, hg must be the identity map on E_2, and h is an isomorphism between (E_1, p_1) and (E_2, p_2). □

Notation: It is often necessary to make the statement "f is a function from space X to space Y which maps a particular point x_0 in X to the point y_0 in

Y." We shall shorten this cumbersome expression by referring to f as a function from the "pair" (X, x_0) to the pair (Y, y_0) and writing $f: (X, x_0) \to (Y, y_0)$.

Minor modifications in the proof of Theorem 5.9 establish the following result. Details of the proof are left as an exercise.

Theorem 5.10. *Let E, B, and X be spaces with base points e_0, b_0, and x_0 respectively, and suppose that (E, p) is a covering space of B with $p(e_0) = b_0$. If $f: (X, x_0) \to (B, b_0)$ is a continuous map for which*

$$f_*\pi_1(X, x_0) \subset p_*\pi_1(E, e_0),$$

then there is a continuous map $\tilde{f}: (X, x_0) \to (E, e_0)$ for which $p\tilde{f} = f$.

In proving Theorem 5.10, keep in mind our agreement that all spaces considered in this chapter are path connected and locally path connected. Actually, Theorem 5.10 remains valid if the requirement on X is reduced to connectedness.

Let us return to our original examples of covering spaces to find the conjugacy class determined by each one. Note that the fundamental group of each base space in these examples is abelian, so each conjugacy class has only one member.

Example 5.8. For the covering $(\mathbb{R}, p)$ over S^1, the fundamental group of $\mathbb{R}$ is trivial so

$$p_{1*}\pi_1(\mathbb{R}) = \{0\},$$

and the conjugacy class consists of only the trivial subgroup of $\pi_1(S^1)$.

Example 5.9. The map $q_n: S^1 \to S^1$ defined by

$$q_n(z) = z^n, \qquad z \in S^1,$$

wraps S^1 around itself n times. Thus if $[\alpha] \in \pi_1(S^1)$, the loop $q_n\alpha$ has degree

$$\deg(q_n\alpha) = n \deg \alpha.$$

Representing $\pi_1(S^1)$ as the group of integers, it follows that $q_{n*}\pi_1(S^1, 1)$ is the subgroup of $\mathbb{Z}$ consisting of all multiples of the integer n.

Example 5.10. If $i: X \to X$ is the identity map, then

$$i_*\pi_1(X) = \pi_1(X),$$

so the conjugacy class in this case contains only the fundamental group of X.

Example 5.11. Consider the double covering (S^2, p) over the projective plane P. The 2-sphere is simply connected, so the conjugacy class contains only the trivial subgroup.

Example 5.12. The plane is simply connected, so the conjugacy class of $(\mathbb{R}^2, r)$ over the torus also contains only the trivial subgroup.

Example 5.13. The infinite spiral Q is contractible and thus has trivial fundamental group. Then (Q, q) determines the conjugacy class of $\pi_1(S^1)$ consisting of only the trivial subgroup. This is the conjugacy class determined in Example 5.8, so Theorem 5.9 shows that (Q, q) and $(\mathbb{R}, p)$ are isomorphic covering spaces of S^1.

The only subgroups of $\pi_1(S^1) = \mathbb{Z}$ are the groups W_n of all multiples of the non-negative integer n. Since $\mathbb{Z}$ is abelian, each singleton set $\{W_n\}$ is a conjugacy class. The subgroup $W_0 = \{0\}$ corresponds to the covering space $(\mathbb{R}, p)$ of Example 5.8, and W_n corresponds to the covering (S^1, q_n) of Example 5.9, $n = 1, 2, \ldots$. By the classification of covering spaces given in Theorems 5.8 and 5.9, any covering space of S^1 must be isomorphic either to $(\mathbb{R}, p)$ or to one of the coverings (S^1, q_n). The next section and the exercises at the end of the chapter provide additional examples of base spaces for which all possible covering spaces can be listed.

5.4 Universal Covering Spaces

If B is a topological space, there is always a covering space corresponding to the conjugacy class of the entire fundamental group, namely (B, i) where i is the identity map on B. This covering space is of little interest for obvious reasons. At the other extreme, the covering space corresponding to the conjugacy class of the trivial subgroup $\{0\}$ of $\pi_1(B)$ is the most interesting. This covering space, if it exists for a particular base space, is called the "universal covering space." This section will examine the relation between a base space B and its universal covering space.

Definition. Let B be a space. A covering space (U, q) of B for which U is simply connected is called the *universal covering space* of B.

The appropriateness of the appellation "the universal covering space" is explained by the next theorem.

Theorem 5.11. (a) *Any two universal covering spaces of a base space B are isomorphic.*

(b) *If (U, q) is the universal covering space of B and (E, p) is a covering space of B, then there is a continuous map $r\colon U \to E$ such that (U, r) is a covering space of E.*

Proof. Statement (a) follows immediately from Theorem 5.9 since any universal covering space determines the conjugacy class of the trivial subgroup.

For part (b), consider the diagram

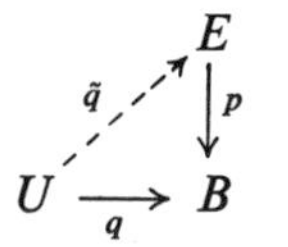

and choose base points u_0, e_0, and b_0 in U, E, and B respectively for which

$$q(u_0) = p(e_0) = b_0.$$

Since $\pi_1(U)$ is trivial, then

$$q_*\pi_1(U, u_0) \subset p_*\pi_1(E, e_0),$$

and Theorem 5.10 guarantees the existence of a continuous map $\tilde{q}: (U, q_0) \to (E, e_0)$ for which $p\tilde{q} = q$. This means that $r = \tilde{q}$ is a covering space homomorphism, and therefore a covering projection, for U over E. □

Definition. Let (E, p) be a covering space of B. An isomorphism from (E, p) to itself is called an *automorphism*. Under the operation of composition, the set of automorphisms of (E, p) forms a group. This group is called the *group of automorphisms* of (E, p) and is denoted by $A(E, p)$.

Proofs of the following remarks are left as exercises:

(a) If f and g are automorphisms of (E, p) and $f(x) = g(x)$ for some x, then $f = g$.
(b) The only member of $A(E, p)$ that has a fixed point is the identity map.

Theorem 5.12. *If (U, q) is the universal covering space of B, then $A(U, q)$ is isomorphic to $\pi_1(B)$. The order of $\pi_1(B)$ is the number of sheets of the universal covering space.*

Proof. Choose a base point b_0 in B and a point u_0 in U for which $q(u_0) = b_0$. We shall first define a function $T: A(U, q) \to \pi_1(B)$.

For $f \in A(U, q)$, $f(u_0)$ is a point in U. Let γ be a path in U from u_0 to $f(u_0)$. Since $qf = q$, then $f(u_0) \in q^{-1}(b_0)$, and hence $q\gamma$ is a loop in B with base point b_0. We thus define T by

$$T(f) = [q\gamma], \qquad f \in A(U, q).$$

Since U is simply connected, the choice of path γ from u_0 to $f(u_0)$ does not affect the homotopy class $[q\gamma]$. Thus T is well-defined.

To see that T is a homomorphism, let $f_1, f_2 \in A(U, q)$ and let γ_1, γ_2 denote paths in U from u_0 to $f_1(u_0)$ and $f_2(u_0)$ respectively. Then

$$T(f_1) = [q\gamma_1], \qquad T(f_2) = [q\gamma_2].$$

The product path $\gamma_1 * f_1\gamma_2$ is a path from u_0 to $f_1f_2(u_0)$. Thus

$$\begin{aligned} T(f_1f_2) &= [q(\gamma_1 * f_1\gamma_2)] = [q\gamma_1 * qf_1\gamma_2] = [q\gamma_1 * q\gamma_2] \\ &= [q\gamma_1] \circ [q\gamma_2] = T(f_1) \circ T(f_2), \end{aligned}$$

so T is a homomorphism.

To see that T is one-to-one, suppose that $T(f_1) = T(f_2)$. Thus the loops $q\gamma_1$ and $q\gamma_2$ determined by f_1 and f_2 are equivalent. The Monodromy Theorem (Theorem 5.5) then implies that $f_1(u_0) = f_2(u_0)$. Thus $f_1 = f_2$, since distinct automorphisms must disagree at every point.

It remains to be shown that T maps $A(U, q)$ onto $\pi_1(B, b_0)$. Let $[\alpha] \in \pi_1(B, b_0)$, and let $\tilde{\alpha}$ denote the unique covering path of α beginning at u_0. Since U is simply connected, we can apply Theorem 5.10 to the diagram

$$\begin{array}{ccc} & & (U, \tilde{\alpha}(1)) \\ & {\scriptstyle h} \nearrow & \downarrow {\scriptstyle q} \\ (U, u_0) & \xrightarrow[q]{} & (B, b_0) \end{array}$$

to obtain a continuous lifting h of q such that $h(u_0) = \tilde{\alpha}(1)$. Since commutativity of the diagram requires $qh = q$, then h is a homomorphism. Reversing the roles of $\tilde{\alpha}(1)$ and u_0 determines a homomorphism k on (U, q) such that $k(\tilde{\alpha}(1)) = u_0$. But then hk and kh are the identity map on U since they are homomorphisms which agree with the identity at some point. Thus $k = h^{-1}$, h is an automorphism, and

$$T(h) = [q\tilde{\alpha}] = [\alpha].$$

This completes the proof that $A(U, q)$ and $\pi_1(B)$ are isomorphic.

The proof that the order of $\pi_1(B)$ is the number of sheets of the universal covering space can be gleaned from what has already been done. The fact that T is one-to-one establishes a one-to-one correspondence between $q^{-1}(b_0)$ and a subset of $\pi_1(B, b_0)$. In proving that T is onto, we showed that every homotopy class $[\alpha]$ in $\pi_1(B, b_0)$ corresponds to a point $\tilde{\alpha}(1)$ in $q^{-1}(b_0)$. Thus the cardinal number of $q^{-1}(b_0)$, which is the number of sheets of (U, q), must equal the order of $\pi_1(B)$. □

The real line is simply connected, so the covering space $(\mathbb{R}, p)$ of Example 5.8 is the universal covering space of the unit circle. Since the plane is simply connected, then the covering space $(\mathbb{R}^2, r)$ of Example 5.12 is the universal covering space of the torus.

Example 5.14. Consider the double covering (S^2, p) of the projective plane P defined in Example 5.4. Since $\pi_1(S^2) = \{0\}$, then (S^2, p) is the universal covering space of P. Moreover, Theorem 5.12 allows us to determine $\pi_1(P)$ by determining $A(S^2, p)$. Since p identifies pairs of antipodal points, then (S^2, p) has two automorphisms, the identity map and the antipodal map. Thus $A(S^2, p)$ is the cyclic group of order two, and $\pi_1(P)$ is the same group. Thus $\pi_1(P)$ is essentially the group of integers modulo 2.

This example generalizes to higher dimensions as follows:

Definition. Let P^n denote the quotient space of the n-sphere S^n obtained by identifying each pair of antipodal points x and $-x$. Then P^n is called *projective n-space.*

The quotient map $p: S^n \to P^n$ is a covering projection. By repeating the reasoning of Example 5.14, the reader can show that the fundamental group of each projective space P^n, $n \geq 2$, is isomorphic to the group of integers modulo 2. A moment's reflection will show that P^1 is homeomorphic to S^1 and hence that $\pi_1(P^1)$ is not the group of integers mod 2.

The classification of covering spaces given in Theorem 5.9 shows that two covering spaces of a space B are isomorphic if and only if they determine the same conjugacy class of subgroups of $\pi_1(B)$. This leaves open the question of the existence of covering spaces. Given a conjugacy class in $\pi_1(B)$, is there a covering space that determines this class? In particular, does every space have a universal covering space? The answer is negative for both questions. Two of the exercises for this chapter give examples of spaces that have no universal covering space. Necessary and sufficient conditions for the existence of a universal covering space are known, but presenting them would take us rather far afield. Readers interested in pursuing this topic should consult references [16] and [20].

5.5 Applications

This section gives two illustrations of the interplay between covering spaces and fundamental groups. The first elucidates the structure of a particular fundamental group, and the second proves part of the famous Borsuk–Ulam Theorem.

Example 5.15. Thus far, all our examples of fundamental groups have been abelian. We shall use covering spaces to provide an example of a nonabelian one.

Let the base space B consist of two tangent circles,

$$B = \{(z, w) \in S^1 \times S^1 : z = 1 \text{ or } w = 1\},$$

and let

$$E = \{(x, y) \in \mathbb{R}^2 : x \text{ or } y \text{ is an integer}\}.$$

Then the map $p: E \to B$ defined by

$$p(x, y) = (e^{2\pi i x}, e^{2\pi i y}), \qquad (x, y) \in \mathbb{R}^2,$$

is a covering projection. Referring to Figure 5.6, p maps each horizontal segment of a square of E once around the left hand circle and each vertical segment of a square of E once around the right hand circle of B.

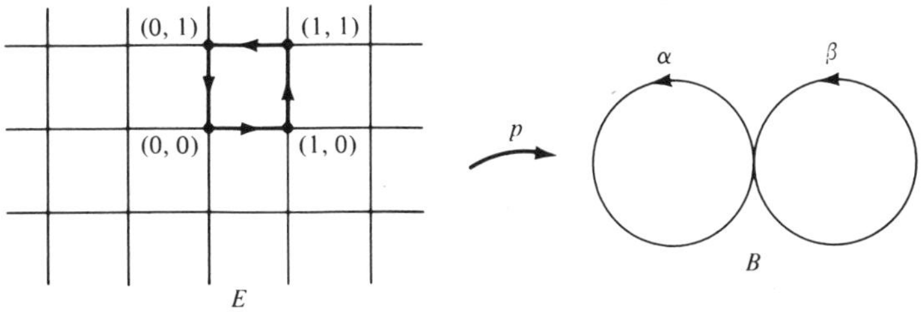

Figure 5.6

Let γ denote the loop in E based at (0, 0) indicated by the arrows, and let $[\alpha]$ and $[\beta]$ denote generators of the fundamental groups of the left and right circles of B respectively. Then $[\gamma]$ is not the identity of $\pi_1(E)$, so

$$p_*([\gamma]) = [\alpha] \circ [\beta] \circ [\alpha]^{-1} \circ [\beta]^{-1}$$

is not the identity in $\pi_1(B)$ since p_* is one-to-one (Theorem 5.7). But if $\pi_1(B)$ were abelian, the commutator $[\alpha] \circ [\beta] \circ [\alpha]^{-1} \circ [\beta]^{-1}$ would be the identity element of $\pi_1(B)$. Thus $\pi_1(B)$ is not abelian. Those readers familiar with free groups may want to prove that $\pi_1(B)$ is the free group generated by $[\alpha]$ and $[\beta]$.

The following theorem was conjectured by S. Ulam and proved by K. Borsuk in 1933:

Theorem 5.13 (The Borsuk–Ulam Theorem). *There is no continuous map $f: S^n \to S^{n-1}$ for which $f(-x) = -f(x)$ for all $x \in S^n$, $n \geq 1$.*

The theorem states that there is no continuous map from S^n to a sphere of lower dimension which maps antipodal points to antipodal points. Such a map would be said to "preserve antipodal points" and would be called "antipode preserving." Since S^0 is a discrete space of two points and therefore not connected, the result is clear for the case $n = 1$. We shall use a covering space argument for the case $n = 2$. A proof for the entire theorem can be found in [20].

Proceeding with the case $n = 2$ by contradiction, suppose that $f: S^2 \to S^1$ is a continuous map for which $f(-x) = -f(x)$ for all $x \in S^2$. Consider the diagram

$$\begin{array}{ccc} S^2 & \xrightarrow{f} & S^1 \\ \downarrow{\scriptstyle p} & & \downarrow{\scriptstyle q} \\ P^2 & \overset{h}{\dashrightarrow} & P^1 \end{array}$$

where (S^2, p) and (S^1, q) denote the double coverings of the projective spaces P^2 and P^1. Even though p^{-1} is not single valued, the fact that f preserves antipodal points guarantees that

$$h = qfp^{-1}: P^2 \to P^1$$

is well-defined and continuous. Note also that the diagram is commutative. Since $\pi_1(P^2)$ is cyclic of order 2 and $\pi_1(P^1) \cong \pi_1(S^1)$ is infinite and cyclic, the induced homomorphism

$$h_*: \pi_1(P^2) \to \pi_1(P^1)$$

must be trivial. Let y_0 be a point of S^2, and let $b_0 = qf(y_0)$ be the base point of P^1. If α is a path in S^2 from y_0 to $-y_0$, then $qf\alpha$ is a loop in P^1. This loop is not equivalent to the constant loop c at b_0 for the following reason: If

$qf\alpha \sim_{b_0} c$, the Monodromy Theorem (Theorem 5.5) guarantees that $f\alpha$ is equivalent to the constant loop based at $f(y_0)$. Since f preserves antipodal points, then

$$f\alpha(1) = f(-y_0) = -f(y_0),$$

so $f\alpha$ is not a loop, and hence cannot possibly be equivalent to a loop. Thus

$$[qf\alpha] \neq [c].$$

Then

$$h_*([p\alpha]) = [hp\alpha] = [qfp^{-1}p\alpha] = [qf\alpha]$$

is not the identity of $\pi_1(B, b_0)$, and h_* is not the trivial homomorphism. This is a contradiction showing that our original assumption that such a map as f exists must be false.

Corollary 1. *Let $g\colon S^2 \to \mathbb{R}^2$ be a continuous map such that $g(-x) = -g(x)$ for all x in S^2. Then $g(x) = 0$ for some x in S^2.*

PROOF. Suppose on the contrary that $g(x)$ is never 0. Then the map $f\colon S^2 \to S^1$ defined by

$$f(x) = g(x)/\|g(x)\|, \qquad x \in S^2,$$

contradicts the Borsuk–Ulam Theorem for the case $n = 2$. □

Corollary 2. *Let $h\colon S^2 \to \mathbb{R}^2$ be a continuous map. Then there is at least one pair $x, -x$ of antipodal points for which $h(x) = h(-x)$.*

PROOF. Assume to the contrary that $h(x) = h(-x)$ for no x in S^2. Then the function $g\colon S^2 \to \mathbb{R}^2$ defined by

$$g(x) = h(x) - h(-x), \qquad x \in S^2,$$

contradicts Corollary 1. □

The last corollary has an interesting physical interpretation. Imagine the surface of the earth to be a 2-dimensional sphere, and suppose that the functions $a(x)$ and $t(x)$ which measure the atmospheric pressure and temperature at x are continuous. Then the map $h\colon S^2 \to \mathbb{R}^2$ defined by

$$h(x) = (a(x), t(x)), \qquad x \in S^2,$$

is continuous. Corollary 2 guarantees that there is at least one pair of antipodal points on the surface of the earth having identical atmospheric pressures and identical temperatures!

The theory of covering spaces developed during the late nineteenth and early twentieth centuries from the theory of Riemann surfaces. Covering spaces were studied, in fact, before the introduction of the fundamental group. Poincaré introduced universal covering spaces in 1883 to prove a theorem about analytic functions [53]. He considered the universal covering space (U, q) of a space B to be the "strongest" covering space of B in the following sense: A curve γ in U is closed if and only if for every covering space (E, p)

of B and every curve γ' in E for which $p\gamma' = q\gamma$, γ' is a closed curve. Exercises at the end of the chapter show that this condition is satisfied if U is simply connected and that (U, q) is indeed the "strongest" covering space of B in the sense of Theorem 5.11.

Covering spaces provided the first example of the power of the fundamental group in classifying topological spaces. We have seen in Theorem 5.9 that the fundamental group accomplishes for covering spaces the type of classification that the homology groups provide for closed surfaces (Theorem 2.11). In addition, the theory of covering spaces was the precursor of the general fiber spaces of Witold Hurewicz and J. P. Serre which are crucial in any advanced course in algebraic topology.

We shall not return in this book to the important and difficult problem of determining fundamental groups. Those interested in this problem should proceed to Van Kampen's Theorem which shows that, under the proper conditions, $\pi_1(X)$ can be determined from the fundamental groups of certain subspaces of X. This theorem and related results can be found in [16] and [19].

Exercises

1. (a) Give an example of a space that is path connected but not locally path connected.
(b) Give an example of a space that is locally path connected but not path connected.

2. Prove that a space X is locally path connected if and only if each path component of each open subset of X is open.

3. Is each component of a space contained in a path component, or is it the other away around? Prove your answer, and give an example to show that components and path components may not be identical.

4. Show that the projection of a "hairpin" onto an interval, as indicated in Figure 5.7, is not a covering projection.

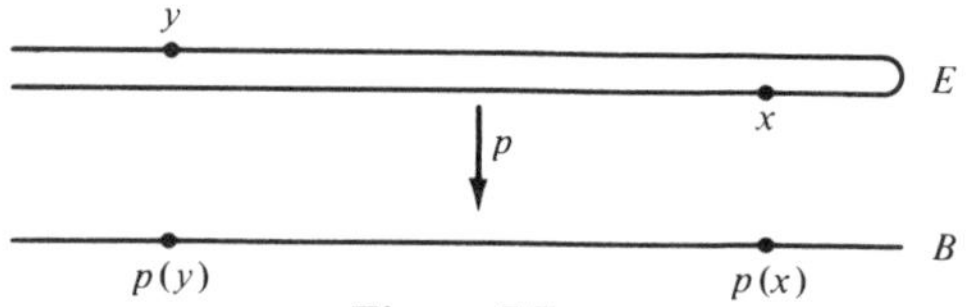

Figure 5.7

5. Definition. A function $f: X \to Y$ is a *local homeomorphism* provided that each point x in X has an open neighborhood U such that f maps U homeomorphically onto $f(U)$.

(a) Prove that every covering projection is a local homeomorphism.
(b) Give an example to show that a local homeomorphism may fail to be a covering projection.

6. Let (E, p) be a covering space of B. Show that the family of admissible neighborhoods is a basis for the topology of B.

7. Repeat Example 5.2 in the case n is a negative integer.

8. Prove the Covering Homotopy Property (Theorem 5.4).

9. Prove the following generalizations of the Covering Homotopy Property:
   (a) **Theorem.** *Let* (E, p) *be a covering space of* B, X *a simply connected space,* $f\colon X \to E$ *a continuous map, and* $H\colon X \times I \to B$ *a homotopy such that* $H(\cdot, 0) = pf$. *Then there is a covering homotopy* $\tilde{H}\colon X \times I \to E$ *of* H *such that* $\tilde{H}(\cdot, 0) = f$.
   (b) Prove the preceding theorem under the assumption that X is a compact Hausdorff space that is not necessarily simply connected.

10. Complete the details in the proof of Theorem 5.6.

11. Prove Theorem 5.7.

12. Prove that a homomorphism of covering spaces is a covering projection.

13. Show that isomorphism of covering spaces is an equivalence relation.

14. Complete the proof of Theorem 5.9.

15. Prove Theorem 5.10.

16. Determine all covering spaces of the torus and exhibit a representative covering space from each isomorphism class.

17. If B is a simply connected space and (E, p) is a covering space of B, prove that p is a homeomorphism from E onto B.

18. Show that the map $p\colon E \to B$ of Example 5.15 is a covering projection.

19. (a) Prove that the set $A(E, p)$ of all automorphisms of a covering space (E, p) is a group.
    (b) Prove that members f, g of $A(E, p)$ must be identical or must agree at no point of E.
    (c) Prove that the identity map is the only member of $A(E, p)$ that has a fixed point.

20. Prove that if B is simply connected, then (B, i) is the universal covering space of B. (Here i denotes the identity map.)

21. Prove that the fundamental group $\pi_1(P^n)$ of projective n-space P^n is isomorphic to the group of integers modulo 2 for $n \geq 2$. What about $n = 1$?

22. Prove that any continuous map $f\colon P^n \to S^1$, $n \geq 2$, from projective n-space to the unit circle is null-homotopic.

23. If (E, p) is a covering space of B and (F, q) is a covering space of C, prove that $(E \times F, p \times q)$ is a covering space of $B \times C$, where $p \times q$ denotes the natural product map.

24. Use Theorem 5.12 to prove that $\pi_1(S^1) \cong \mathbb{Z}$ and $\pi_1(S^1 \times S^1) \cong \mathbb{Z} \oplus \mathbb{Z}$.

25. Let G and $\tilde{G}$ be path connected and locally path connected topological groups and $p\colon \tilde{G} \to G$ a group homomorphism for which $(\tilde{G}, p)$ is a covering space of G. Prove that the kernel of p is isomorphic to $A(\tilde{G}, p)$.

26. Prove that an infinite product of circles has no universal covering space.

27. Let X be the subset of the plane consisting of the circumferences of circles having radius $1/n$ and center at $(1/n, 0)$ for $n = 1, 2, \ldots$. Show that X has no universal covering space.

28. Let (E, p) be a covering space of B, and let e_0, b_0 be points of E and B respectively with $p(e_0) = b_0$.

(a) Show that there is a one-to-one correspondence between $p^{-1}(b_0)$ and the set of left cosets $\pi_1(B, b_0)/p_*\pi_1(E, e_0)$.

(b) **Definition.** The covering space (E, p) is called *regular* if $p_*\pi_1(E, e_0)$ is a normal subgroup of $\pi_1(B, b_0)$.

Show that regularity is not dependent on the choice of base point e_0 in $p^{-1}(b_0)$. (*Hint*: Use conjugacy classes.)

(c) Prove that the automorphism group $A(E, p)$ is isomorphic to the quotient group $\pi_1(B, b_0)/p_*\pi_1(E, e_0)$ if (E, p) is regular. Deduce Theorem 5.12 as a corollary.

29. Let us say that a covering space (U, q) of B *satisfies Property P* if it is the "strongest" covering space of B in the sense of Poincaré: A curve γ in U is closed if and only if for every covering space (E, p) of B and every curve γ' in E for which $p\gamma' = q\gamma$, γ' is a closed curve.

Prove:

(a) If U is simply connected, then (U, q) satisfies Property P.

(b) Any two covering spaces of B which satisfy Property P are isomorphic.

(c) If (U, q) satisfies Property P and (E, p) is any covering space of B, then there is a homomorphism $r: U \to E$ for which (U, r) is a covering space of E.

6 The Higher Homotopy Groups

6.1 Introduction

The fundamental group of a connected polyhedron provides more information than does its first homology group. This is evident from Theorem 4.11 since the first homology group is completely determined by the fundamental group. For this reason, the need for higher dimensional analogues of the fundamental group was recognized early in the development of algebraic topology. Definitions of these "higher homotopy groups" were given in the years 1932–1935 by Eduard Čech (1893–1960) and Witold Hurewicz (1904–1956). It was Hurewicz who gave the most satisfactory definition and proved the fundamental properties.

Let us consider in an intuitive way the possible methods of defining the second homotopy group $\pi_2(X, x_0)$ of a space X at a point x_0 in X. Recall that $\pi_1(X, x_0)$ is the set of homotopy classes of loops in X based at x_0. Our first problem is to define what one might call a "2-dimensional loop."

A "1-dimensional loop" is a continuous map $\alpha: I \to X$ for which the boundary points 0 and 1 have image x_0. We might then define a 2-dimensional loop to be a continuous map $\beta: I \times I \to X$ from the unit square into X which maps the boundary of the square to x_0.

From a slightly different point of view, we can consider a loop α in X as a continuous map from S^1 to X which takes 1 to x_0. This follows from the observation that the quotient space of the unit interval I obtained by identifying 0 and 1 to a single point is simply S^1. Thus another possible definition of 2-dimensional loop is a continuous map from the 2-sphere S^2 into X. Note that both of these definitions of 2-dimensional loop generalize to higher dimensions by considering higher dimensional cubes and spheres.

There is a third possibility. Perhaps a 2-dimensional loop should be a "loop of loops." That is to say, perhaps a 2-dimensional loop should be a function β having domain I with each value $\beta(t)$ a loop in X, and having the additional property $\beta(0) = \beta(1)$. This idea is the point of genius in Hurewicz' approach. Carrying it out will involve defining a topology for the set $\Omega(X, x_0)$ of loops in X with base point x_0. Once this topology is determined, one can define $\pi_2(X, x_0)$ to be the fundamental group of $\Omega(X, x_0)$.

It is remarkable that all three approaches lead to the same group $\pi_2(X, x_0)$. The next section presents the definitions based on these three ideas and shows that the same group is determined in each case.

6.2 Equivalent Definitions of $\pi_n(X, x_0)$

We shall take the three definitions in the order in which they have been discussed. If n is a positive integer, the symbol I^n denotes the *unit n-cube*

$$I^n = \{t = (t_1, t_2, \ldots, t_n) \in \mathbb{R}^n : 0 \le t_i \le 1 \text{ for each } i\}$$

and ∂I^n, called the *boundary* of I^n, denotes its point set boundary

$$\partial I^n = \{t = (t_1, t_2, \ldots, t_n) \in I^n : \text{some } t_i \text{ is } 0 \text{ or } 1\}.$$

Note that the boundary symbol ∂ must not be confused with the boundary operator of homology theory.

Definition A. Let X be a space and x_0 a point of X. For a given positive integer n, consider the set $F_n(X, x_0)$ of all continuous maps α from the unit n-cube I^n into X for which $\alpha(\partial I^n) = x_0$. Define an equivalence relation $\sim_{x_0}$ on $F_n(X, x_0)$ as follows: For α and β in $F_n(X, x_0)$, α is *equivalent modulo* x_0 *to* β, written $\alpha \sim_{x_0} \beta$, if there is a homotopy $H: I^n \times I \to X$ such that

$$H(t_1, \ldots, t_n, 0) = \alpha(t_1, \ldots, t_n),$$
$$H(t_1, \ldots, t_n, 1) = \beta(t_1, \ldots, t_n), \qquad (t_1, \ldots, t_n) \in I^n,$$

and

$$H(t_1, \ldots, t_n, s) = x_0, \qquad (t_1, \ldots, t_n) \in \partial I^n, \ s \in I.$$

In shorter form the requirements on the homotopy H are

$$H(\cdot, 0) = \alpha, \qquad H(\cdot, 1) = \beta,$$
$$H(\partial I^n \times I) = x_0.$$

Under this equivalence relation on $F_n(X, x_0)$, the equivalence class determined by α is denoted $[\alpha]$ and called the *homotopy class of* α *modulo* x_0 or simply the *homotopy class of* α.

Define an operation $*$ on $F_n(X, x_0)$ as follows: For α, β in $F_n(X, x_0)$,

$$\alpha * \beta(t_1, \ldots, t_n) = \begin{cases} \alpha(2t_1, t_2, \ldots, t_n) & \text{if } 0 \le t_1 \le \frac{1}{2} \\ \beta(2t_1 - 1, t_2, \ldots, t_n) & \text{if } \frac{1}{2} \le t_1 \le 1. \end{cases}$$

Note that the $*$ operation is completely determined by the first coordinate of the variable point $(t_1, \ldots, t_n)$ and that the continuity of $\alpha * \beta$ follows from the Continuity Lemma. The $*$ operation induces an operation $\circ$ on the set of homotopy classes of $F_n(X, x_0)$:

$$[\alpha] \circ [\beta] = [\alpha * \beta].$$

With this operation, the set of equivalence classes of $F_n(X, x_0)$ is a group. This group is called the nth *homotopy group* of X at x_0 and is denoted by $\pi_n(X, x_0)$.

As in the case of the fundamental group, the definition requires that some details be verified:

(1) The relation $\sim_{x_0}$ is an equivalence relation on $F_n(X, x_0)$.
(2) The operation $*$ determines the operation $\circ$ completely. In other words, if $\alpha \sim_{x_0} \alpha'$ and $\beta \sim_{x_0} \beta'$, then $\alpha * \beta \sim_{x_0} \alpha' * \beta'$.
(3) With the $\circ$ operation, $\pi_n(X, x_0)$ is actually a group. Its identity is the class $[c]$ determined by the constant map $c(I^n) = x_0$. The inverse $[\alpha]^{-1}$ of $[\alpha]$ is the class $[\bar{\alpha}]$ where $\bar{\alpha}$, called the *reverse* of α, is defined by

$$\bar{\alpha}(t_1, t_2, \ldots, t_n) = \alpha(1 - t_1, t_2, \ldots, t_n), \qquad (t_1, t_2, \ldots, t_n) \in I^n.$$

Since the definition of $\pi_n(X, x_0)$ is completely analogous to that of $\pi_1(X, x_0)$ except for the extra coordinates, the proofs of these details are left as exercises.

The quotient space of I^n obtained by identifying ∂I^n to a point is homeomorphic to the n-sphere S^n. Let us assume that the point of identification is the point $1 = (1, 0, \ldots, 0)$ of S^n having first coordinate unity and all other coordinates zero. Then $\pi_n(X, x_0)$ can be defined in terms of maps from $(S^n, 1)$ to (X, x_0) as follows:

Definition B. For a given positive integer n, consider the set $G_n(X, x_0)$ of all continuous maps α from S^n to X such that $\alpha(1) = x_0$. Define an equivalence relation on $G_n(X, x_0)$ in the following way: For α, β in $G_n(X, x_0)$, α is *equivalent modulo* x_0 *to* β, written $\alpha \sim_{x_0} \beta$, if there is a homotopy $H: S^n \times I \to X$ such that

$$H(\cdot, 0) = \alpha, \qquad H(\cdot, 1) = \beta,$$
$$H(1, s) = x_0, \qquad s \in I.$$

The equivalence class $[\alpha]$ determined by α is called the *homotopy class of* α. The set of homotopy classes is denoted by $\pi_n(X, x_0)$.

In view of the discussion preceding Definition B, it should be clear that there is a natural one-to-one correspondence between $F_n(X, x_0)$ and $G_n(X, x_0)$ under which a map α in $G_n(X, x_0)$ corresponds to the map $\alpha' = \alpha q$ where q is the map from I^n to S^n which identifies ∂I^n to the point 1. Also, two members α and β in $G_n(X, x_0)$ are equivalent modulo x_0 if and only if their counterparts α' and β' are equivalent in $F_n(X, x_0)$. Thus Definitions A and B give equivalent definitions of the set $\pi_n(X, x_0)$. The elements $[\alpha]$ are usually more easily visualized in terms of Definition B.

The $\circ$ operation for Definition B is defined in terms of the identification of I^n to S^n. Let $\alpha, \beta \in G_n(X, x_0)$. The identification map q takes the sets

$$A = \{(t_1, \ldots, t_n) \in I^n: t_1 \leq \tfrac{1}{2}\},$$
$$B = \{(t_1, \ldots, t_n) \in I^n: t_1 \geq \tfrac{1}{2}\}$$

to hemispheres A' and B' respectively of S^n whose intersection

$$A' \cap B' = q(A \cap B)$$

if homeomorphic to S^{n-1}. Imagine that $A' \cap B'$ is identified to the base point 1 by an identification map r. The resulting space consists of two n-spheres tangent at their common base point as in Figure 6.1. The product $\alpha * \beta$ is now defined by

$$\alpha * \beta(x) = \begin{cases} \alpha r(x) & \text{if } x \in A' \\ \beta r(x) & \text{if } x \in B'. \end{cases}$$

The group operation $\circ$ is defined by

$$[\alpha] \circ [\beta] = [\alpha * \beta].$$

Observe that the operation for Definition B has been designed expressly to show that Definitions A and B describe isomorphic groups.

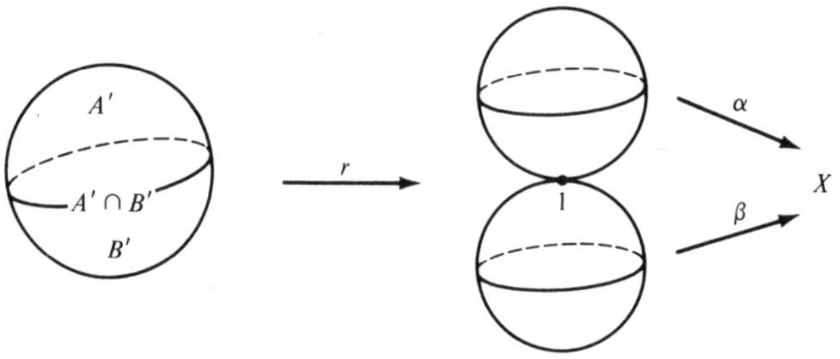

Figure 6.1

The third description of the nth homotopy group requires a topology for the set of loops in X based at x_0.

Definition. Let F be a collection of continuous functions from a space Y into a space Z. If K is a compact subset of Y and U an open subset of Z, let

$$W(K, U) = \{\alpha \in F: \alpha(K) \subset U\}.$$

The family of all such sets $W(K, U)$, as K ranges over the compact sets in Y and U ranges over the open sets in Z, is a subbase for a topology for F. This topology is called the *compact-open topology* for F.

Since we shall apply the compact-open topology only to the set of loops in a space X, we repeat the definition for this case.

Definition. Let X be a space and x_0 a point of X. Consider the set $\Omega(X, x_0)$ of all loops in X with base point x_0. If K is a compact subset of I and U is open in X, let

$$W(K, U) = \{\alpha \in \Omega(X, x_0): \alpha(K) \subset U\}.$$

The family of all such sets $W(K, U)$, where K is compact in I and U is open in X, is a subbase for a topology for $\Omega(x, x_0)$. This topology is the *compact-open topology* for $\Omega(X, x_0)$. Note that basic open sets in the compact-open topology have the form

$$\bigcap_{i=1}^{r} W(K_i, U_i)$$

where $K_1, \ldots, K_r$ are compact sets in I and $U_1, \ldots, U_r$ are open in X. A loop α belongs to this basic open set if and only if $\alpha(K_i) \subset U_i$ for each $i = 1, 2, \ldots, r$.

Theorem 6.1. *If X is a metric space, the compact-open topology for $\Omega(X, x_0)$ is the same as its topology of uniform convergence.*

Proof. Let d denote the metric on X. Recall that the topology of uniform convergence on $\Omega(X, x_0)$ is determined by the metric ρ defined as follows: If α and β are in $\Omega(X, x_0)$, then $\rho(\alpha, \beta)$ is the supremum (or least upper bound) of the distances from $\alpha(t)$ to $\beta(t)$ for t in I:

$$\rho(\alpha, \beta) = \sup\{d(\alpha(t), \beta(t)): t \in I\}.$$

Then the topology of uniform convergence has as a basis the set of all spherical neighborhoods

$$S(\alpha, r) = \{\beta \in \Omega(X, x_0): d(\alpha, \beta) < r\}$$

where $\alpha \in \Omega(X, x_0)$ and r is a positive number.

Let T and T' denote respectively the compact-open topology and the topology of uniform convergence for $\Omega(X, x_0)$. To see that $T \subset T'$, let $W(K, U)$ be a subbasic open set in T, where K is compact in I and U is open in X. Let $\alpha \in W(K, U)$. Since the compact set $\alpha(K)$ is contained in U, there is a positive number ϵ such that any point of X at a distance less than ϵ from $\alpha(K)$ is also in U. Consider the basic open set $S(\alpha, \epsilon)$ in T'. If $\beta \in S(\alpha, \epsilon)$, then for each t in K, $d(\alpha(t), \beta(t)) < \epsilon$. Thus $\beta(t)$ must be in U since its distance from a point of $\alpha(K)$ is less than ϵ. Hence $\beta(K) \subset U$, so $\beta \in W(K, U)$. We now have

$$\alpha \in S(\alpha, \epsilon) \subset W(K, U),$$

so $W(K, U)$ must be open in T'. Then $T \subset T'$ since T' contains a subbase for T.

To see that $T' \subset T$, let $S(\gamma, r)$ with center γ and radius $r > 0$ be a basic open set in T'. To prove that $S(\gamma, r)$ is in T, it is sufficient to find a member of T which contains γ and is contained in $S(\gamma, r)$. (Why?) Let $\{U_j\}$ be a cover of

X by open sets having diameters less than r, and let η be a Lebesgue number for the open cover $\{\gamma^{-1}(U_j)\}$ of I. Let

$$0 = t_0 < t_1 < \cdots < t_n = 1$$

be a subdivision of I with successive points differing by less than η. Then for $i = 1, 2, \ldots, n$, γ maps each of the compact sets $K_i = [t_{i-1}, t_i]$ into one of the open sets of the cover $\{U_j\}$. Choose such an open set, say U_i, for each i so that

$$\gamma(K_i) \subset U_i, \qquad i = 1, 2, \ldots, n.$$

Then

$$\gamma \in \bigcap_{i=1}^{n} W(K_i, U_i),$$

and this set is open in T. If $\beta \in \bigcap_{i=1}^{n} W(K_i, U_i)$, then $\rho(\gamma, \beta)$ cannot exceed the maximum of the diameters of $U_1, \ldots, U_n$. Thus $\rho(\gamma, \beta) < r$, so $\beta \in S(\gamma, r)$. Then $S(\gamma, r)$ is open in T, and T contains T' since it contains a basis for T'. Since it has been shown that $T \subset T'$ and $T' \subset T$, then $T = T'$. □

Definition C. Let X be a space with $x_0 \in X$, and consider the set $\Omega(X, x_0)$ of loops in X based at x_0 with the compact-open topology. If $n \geq 2$, the nth *homotopy group of X at x_0* is the $(n-1)$th homotopy group of $\Omega(X, x_0)$ at c, where c is the constant loop at x_0. Thus

$$\pi_2(X, x_0) = \pi_1(\Omega(X, x_0), c), \ldots$$
$$\pi_n(X, x_0) = \pi_{n-1}(\Omega(X, x_0), c).$$

Definition C for the higher homotopy groups was given by Witold Hurewicz in 1935. His definition was originally applied only to metric spaces, and $\Omega(X, x_0)$ was assigned the topology of uniform convergence. The compact-open topology, which permitted the extension of Hurewicz' definition to arbitrary spaces, was introduced by R. H. Fox (1913–1973) in 1944. The inductive definition expresses each homotopy group ultimately as a fundamental group of a space of loops. This will be helpful in our applications later. This definition has one obvious disadvantage, however. It does not lend itself easily to intuitive considerations. How, for example, can one imagine $\pi_3(X, x_0)$ as the fundamental group of the iterated loop space of X?

Each of the three definitions of the higher homotopy groups has advantages and shortcomings. To understand homotopy theory, one must know all three and must be able to apply the one that fits best in a given situation.

The three definitions A, B, and C of the higher homotopy groups are all equivalent. We have discussed the equivalence of A and B and now turn to a comparison of A with C. This discussion will be for the case $n = 2$ since the extension to higher values of n requires little more than writing additional coordinates.

Suppose then that α is a member of $F_2(X, x_0)$; i.e., α is a continuous map from the unit square I^2 to X which takes ∂I^2 to x_0. Then α determines a member $\hat{\alpha}$ of $\Omega(\Omega(X, x_0), c)$ defined by

$$\hat{\alpha}(t_1)(t_2) = \alpha(t_1, t_2), \qquad t_1, t_2 \in I.$$

Each value $\hat{\alpha}(t_1)$ is a continuous function from I into X because α is continuous. Note that

$$\hat{\alpha}(t_1)(0) = \hat{\alpha}(t_1)(1) = x_0$$

since $(t_1, 0)$ and $(t_1, 1)$ are in ∂I^2. Thus $\hat{\alpha}(t_1) \in \Omega(X, x_0)$, and obviously $\hat{\alpha}(0) = \hat{\alpha}(1)$ is the constant loop c whose only value is x_0. But is $\hat{\alpha}$ continuous as a function from I into $\Omega(X, x_0)$? To see that it is, let $W(K, U)$ be a subbasic open set in $\Omega(X, x_0)$. As usual, K is compact in I and U is open in X. Let $t_1 \in \hat{\alpha}^{-1}(W(K, U))$. Then

$$\hat{\alpha}(t_1)(K) = \alpha(\{t_1\} \times K) \subset U.$$

Since K is compact, there is an open set O in I such that $t_1 \in O$ and

$$\alpha(O \times K) \subset U.$$

Thus

$$t_1 \in O \subset \hat{\alpha}^{-1}(W(K, U)),$$

so $\hat{\alpha}^{-1}(W(K, U))$ is an open set and $\hat{\alpha}$ is continuous. Thus each member of $F_2(X, x_0)$ determines in a natural way a member of $\Omega(\Omega(X, x_0), c)$.

Suppose that we reverse the process and begin with a member $\hat{\alpha}$ of $\Omega(\Omega(X, x_0), c)$. Then $\hat{\alpha}$ determines a function $\alpha: I^2 \to X$ defined by

$$\alpha(t_1, t_2) = \hat{\alpha}(t_1)(t_2), \qquad (t_1, t_2) \in I^2.$$

It is an easy exercise to see that $\alpha \in F_2(X, x_0)$. We have thus established a one-to-one correspondence between $F_2(X, x_0)$ and $\Omega(\Omega(X, x_0), c)$.

Suppose that $H: I^2 \times I \to X$ is a homotopy demonstrating the equivalence of α and β as prescribed in Definition A. Then the homotopy

$$\hat{H}: I \times I \to \Omega(X, x_0)$$

defined by

$$\hat{H}(t_1, s)(t_2) = H(t_1, t_2, s), \qquad t_1, t_2, s \in I,$$

demonstrates the equivalence of the loops $\hat{\alpha}$ and $\hat{\beta}$. Reversing the argument shows that $\hat{\alpha}$ equivalent to $\hat{\beta}$ implies α equivalent to β. Thus there is a one-to-one correspondence between homotopy classes $[\alpha]$ of Definition A and homotopy classes $[\hat{\alpha}]$ of Definition C. Since the $*$ operation in Definition A is completely determined in the first coordinate, it follows that for any $\alpha, \beta \in F_2(X, x_0)$, $[\alpha * \beta]$ corresponds to $[\hat{\alpha} * \hat{\beta}]$ and hence that the two definitions of $\pi_2(X, x_0)$ lead to isomorphic groups.

6.3 Basic Properties and Examples

Many theorems about the fundamental group generalize to the higher homotopy groups. The following three results can be proved by methods very similar to those used to prove their analogues in Chapter 4.

Theorem 6.2. *If the space X is path connected and x_0 and x_1 are points of X, then $\pi_n(X, x_0)$ is isomorphic to $\pi_n(X, x_1)$ for each $n \geq 1$.*

As in the case of the fundamental group, we shall sometimes omit reference to the base point and refer to the "nth homotopy group of X," $\pi_n(X)$, when X is path connected and we are concerned only with the algebraic structure of the group.

Theorem 6.3. *If X is contractible by a homotopy that leaves x_0 fixed, then $\pi_n(X, x_0) = \{0\}$ for each $n \geq 1$.*

Theorem 6.4. *Let X and Y be spaces with points x_0 in X and y_0 in Y. Then*

$$\pi_n(X \times Y, (x_0, y_0)) \cong \pi_n(X, x_0) \oplus \pi_n(Y, y_0), \qquad n \geq 1.$$

Example 6.1. The following spaces are contractible, so each has nth homotopy group $\{0\}$ for each value of n:

(a) the real line,
(b) Euclidean space of any dimension,
(c) an interval,
(d) a convex figure in Euclidean space.

We saw in Chapter 4 that the fundamental group is usually difficult to determine. This is doubly true of the higher homotopy groups. The homotopy groups $\pi_k(S^n)$ of the n-sphere, for example, have never been completely determined. (The hard part is the case $k > n$.) Finding the homotopy groups of S^n is one of the major unsolved problems of algebraic topology. The groups $\pi_k(S^n)$ for $k \leq n$ are computed in the following examples.

Example 6.2. For $k < n$, the kth homotopy group $\pi_k(S^n)$ is the trivial group. To see this, let $[\alpha]$ be a member of $\pi_k(S^n)$, and consider α as a continuous map from $(S^k, 1)$ to $(S^n, 1)$. Represent S^k and S^n as the boundary complexes of simplexes of dimensions $k + 1$ and $n + 1$ respectively. By the Simplicial Approximation Theorem (Theorem 3.6), α has a simplicial approximation $\alpha': S^k \to S^n$ for which $[\alpha] = [\alpha']$. But since a simplicial map cannot map a simplex onto a simplex of higher dimension, then α' is not onto. Let p be a point in S^n which is not in the range of α'. Then $S^n \backslash \{p\}$ is contractible since it is homeomorphic to $\mathbb{R}^n$, and hence α', a map whose range is contained in a contractible space, is null-homotopic. Thus

$$[\alpha] = [\alpha'] = [c],$$

so $\pi_k(S^n)$ is the trivial group whose only member is the class $[c]$ determined by the constant map.

Example 6.3. For $n \geq 1$, the nth homotopy group $\pi_n(S^n)$ is isomorphic to the group $\mathbb{Z}$ of integers. (The case $n = 1$ was considered in some detail in Chapter 4.)

Consider $\pi_n(S^n)$, $n \geq 2$, as the set of homotopy classes of maps $\alpha\colon (S^n, 1) \to (S^n, 1)$ as in Definition B. Define $\rho\colon \pi_n(S^n) \to \mathbb{Z}$ by

$$\rho([\alpha]) = \text{degree of } \alpha, \qquad [\alpha] \in \pi_n(S^n).$$

Brouwer's Degree Theorem (Theorem 3.9) insures that ρ is well-defined, and the Hopf Classification Theorem (Theorem 3.10), which was stated without proof in Chapter 3, shows that it is one-to-one. The identity map $i\colon (S^n, 1) \to (S^n, 1)$ has degree 1, and the description of the $*$ operation in Definition B shows that the map

$$i^k = i * i * \cdots * i \qquad (k \text{ terms})$$

has degree k. Thus $[i]$ is a generator of $\pi_n(S^n)$, and

$$\rho([i]^k) = k, \qquad \rho([i]^{-k}) = -k$$

for any positive integer k. It follows easily that ρ is an isomorphism.

Example 5.15 shows that the fundamental group of a space may fail to be abelian. The higher homotopy groups are all abelian, as we shall see shortly. This property is the result of the $*$ operation in $\Omega(X, x_0)$. The next theorem illustrates the method of proof and serves as an introduction to the more complicated proof of the commutativity of $\pi_n(X, x_0)$ for $n \geq 2$.

Theorem 6.5. *Let G be a topological group with identity element e. Then $\pi_1(G, e)$ is abelian.*

Proof. The operation on G induces an operation $\cdot$ on the set $\Omega(G, e)$ of loops in G based at e defined by

$$\alpha \cdot \beta(t) = \alpha(t)\beta(t), \qquad \alpha, \beta \in \Omega(G, e),\ t \in I,$$

where the juxtaposition of $\alpha(t)$ and $\beta(t)$ indicates their product in G. This operation induces an operation $\square$ on $\pi_1(G, e)$:

$$[\alpha] \square [\beta] = [\alpha \cdot \beta], \qquad [\alpha], [\beta] \in \pi_1(G, e).$$

Let c denote the constant loop at e, and let $[\alpha]$ and $[\beta]$ be members of $\pi_1(G, e)$. Observe that

$$(\alpha * c)\cdot(c * \beta)(t) = \begin{cases} \alpha(2t)e = \alpha(2t) & \text{if } 0 \leq t \leq \frac{1}{2} \\ e\beta(2t - 1) = \beta(2t - 1) & \text{if } \frac{1}{2} \leq t \leq 1, \end{cases}$$

$$(c * \alpha)\cdot(\beta * c)(t) = \begin{cases} e\beta(2t) = \beta(2t) & \text{if } 0 \leq t \leq \frac{1}{2} \\ \alpha(2t - 1)e = \alpha(2t - 1) & \text{if } \frac{1}{2} \leq t \leq 1. \end{cases}$$

This gives

$$(\alpha * c)\cdot(c * \beta) = \alpha * \beta, \qquad (c * \alpha)\cdot(\beta * c) = \beta * \alpha.$$

Then

$$\begin{aligned} [\alpha] \circ [\beta] &= [\alpha * \beta] = [(\alpha * c)\cdot(c * \beta)] = [\alpha * c] \square [c * \beta] \\ &= [c * \alpha] \square [\beta * c] = [(c * \alpha)\cdot(\beta * c)] = [\beta * \alpha] = [\beta] \circ [\alpha], \end{aligned}$$

so $\pi_1(G, e)$ is abelian.

Here is an additional curious fact. The operations $\circ$ and $\square$ are precisely equal:

$$[\alpha] \circ [\beta] = [\alpha * \beta] = [(\alpha * c)\cdot(c * \beta)] = [\alpha * c] \,\square\, [c * \beta] = [\alpha] \,\square\, [\beta]. \qquad \square$$

Not all of the group properties were used in the proof of Theorem 6.5. The existence of a multiplication with identity element e is sufficient, and even that assumption can be weakened. The following definition describes the property that makes the proof work.

Definition. An *H-space* or *Hopf space* is a topological space Y with a continuous multiplication (indicated by juxtaposition) and a point y_0 in Y for which the map defined by multiplying on the left by y_0 and the map defined by multiplying on the right by y_0 are both homotopic to the identity map on Y by homotopies that leave y_0 fixed. In other words, there exist homotopies L and R from $Y \times I$ into Y such that

$$L(y, 0) = y_0 y, \qquad L(y, 1) = y, \qquad L(y_0, t) = y_0,$$
$$R(y, 0) = y y_0, \qquad R(y, 1) = y, \qquad R(y_0, t) = y_0$$

for all y in Y and t in I. The point y_0 is called the *homotopy unit* of Y.

Note that any topological group is an H-space. H-spaces were first considered by Heinz Hopf, and they are named in his honor.

Example 6.4. If X is a space and x_0 a point of X, then the loop space $\Omega(X, x_0)$ with the compact-open topology is an H-space. The multiplication is the $*$ operation, and the homotopy unit is the constant map c. The required homotopies L and R are defined for α in $\Omega(X, x_0)$ and s in I by

$$L(\alpha, s)(t) = \begin{cases} x_0 & \text{if } 0 \le t \le (1 - s)/2 \\ \alpha\left(\dfrac{2t + s - 1}{s + 1}\right) & \text{if } (1 - s)/2 \le t \le 1, \end{cases}$$

$$R(\alpha, s)(t) = \begin{cases} \alpha\left(\dfrac{2t}{s + 1}\right) & \text{if } 0 \le t \le (s + 1)/2 \\ x_0 & \text{if } (s + 1)/2 \le t \le 1. \end{cases}$$

The reader is left the exercise of proving that the multiplication $*$ and the homotopies L and R are continuous with respect to the compact-open topology.

Theorem 6.6. *If Y is an H-space with homotopy unit y_0, then $\pi_1(Y, y_0)$ is abelian.*

PROOF. The operation on Y induces an operation $\cdot$ on $\Omega(Y, y_0)$ as in the proof of Theorem 6.5:

$$\alpha\cdot\beta(t) = \alpha(t)\beta(t), \qquad \alpha, \beta \in \Omega(Y, y_0),\ t \in I.$$

This operation likewise induces an operation $\square$ on $\pi_1(Y, y_0)$:

$$[\alpha] \square [\beta] = [\alpha \cdot \beta], \qquad [\alpha], [\beta] \in \pi_1(Y, y_0).$$

Letting c denote the constant loop at y_0,

$$(\alpha * c)\cdot(c * \beta)(t) = \begin{cases} \alpha(2t)y_0 & \text{if } 0 \le t \le \frac{1}{2} \\ y_0\beta(2t - 1) & \text{if } \frac{1}{2} \le t \le 1, \end{cases}$$

$$(c * \alpha)\cdot(\beta * c)(t) = \begin{cases} y_0\beta(2t) & \text{if } 0 \le t \le \frac{1}{2} \\ \alpha(2t - 1)y_0 & \text{if } \frac{1}{2} \le t \le 1. \end{cases}$$

Since multiplication on the left by y_0 and multiplication on the right by y_0 are both homotopic to the identity map on Y, then

$$[(\alpha * c)\cdot(c * \beta)] = [\alpha * \beta],$$
$$[(c * \alpha)\cdot(\beta * c)] = [\beta * \alpha].$$

Thus

$$[\alpha] \circ [\beta] = [\alpha * \beta] = [(\alpha * c)\cdot(c * \beta)] = [(c * \alpha)\cdot(\beta * c)]$$
$$= [\beta * \alpha] = [\beta] \circ [\alpha].$$

It follows as in the proof of Theorem 6.5 that the operations $\circ$ and $\square$ are equal. □

Theorem 6.7. *The higher homotopy groups $\pi_n(X, x_0)$, $n \ge 2$, of any space X are abelian.*

Proof. The second homotopy group

$$\pi_2(X, x_0) = \pi_1(\Omega(X, x_0), c)$$

is abelian since $\Omega(X, x_0)$ is an H-space with the constant loop c as homotopy unit. Proceeding inductively, suppose that the $(n - 1)$th homotopy group $\pi_{n-1}(Y, y_0)$ is abelian for every space Y. Then

$$\pi_n(X, x_0) = \pi_{n-1}(\Omega(X, x_0), c)$$

must be abelian, and the proof is complete. □

Definition. Let $f: (X, x_0) \to (Y, y_0)$ be a continuous map on the indicated pairs. If $[\alpha] \in \pi_n(X, x_0)$, $n \ge 1$, then the composition $f\alpha: I^n \to Y$ is a continuous map which takes ∂I^n to y_0, so $f\alpha$ represents an element $[f\alpha]$ in $\pi_n(Y, y_0)$. Thus f induces a function

$$f_*: \pi_n(X, x_0) \to \pi_n(Y, y_0)$$

defined by

$$f_*([\alpha]) = [f\alpha], \qquad [\alpha] \in \pi_n(X, x_0).$$

The function f_* is called the *homomorphism induced by f* in dimension n.

To be very precise we should refer to f_*^n, indicating the dimension n, but this notation is cumbersome, and we shall avoid it. The dimension in question will always be known from the subscripts on the homotopy groups involved. The reader is left the exercise of showing that f_* is actually a well-defined homomorphism.

Theorem 6.8. (a) *If* $f: (X, x_0) \to (Y, y_0)$ *and* $g: (Y, y_0) \to (Z, z_0)$ *are continuous maps on the indicated pairs, then the induced homomorphism* $(gf)_*$ *is the composite map*

$$g_* f_*: \pi_n(X, x_0) \to \pi_n(Z, z_0)$$

in each dimension n.

(b) *If* $h: (X, x_0) \to (Y, y_0)$ *is a homeomorphism, then the homomorphism* h_* *induced by h is an isomorphism for each value of n.*

PROOF. (a) If $[\alpha] \in \pi_n(X, x_0)$, then

$$(fg)_*([\alpha]) = [gf\alpha] = g_*([f\alpha]) = g_* f_*([\alpha]),$$

so

$$(gf)_* = g_* f_*.$$

(b) Suppose that $h^{-1}: (Y, y_0) \to (X, x_0)$ is the inverse of h. Then for $[\alpha]$ in $\pi_n(X, x_0)$,

$$(h^{-1})_* h_*([\alpha]) = [h^{-1}h\alpha] = [\alpha],$$

so $(h^{-1})_* h_*$ is the identity map on $\pi_n(X, x_0)$. By symmetry it follows that $h_*(h^{-1})_*$ is the identity map on $\pi_n(Y, y_0)$, so h_* is an isomorphism. □

It was proved in Chapter 5 that a covering projection $p: E \to B$ induces a monomorphism (i.e., a one-to-one homomorphism) $p_*: \pi_1(E) \to \pi_1(B)$. The next theorem, discovered by Hurewicz, shows that the induced homomorphism for the higher homotopy groups is even better.

Theorem 6.9. *Let* (E, p) *be a covering space of B, and let* e_0 *in E and* b_0 *in B be points such that* $p(e_0) = b_0$. *Then the induced homomorphism*

$$p_*: \pi_n(E, e_0) \to \pi_n(B, b_0)$$

is an isomorphism for $n \geq 2$.

PROOF. To see that p_* is onto, consider an element $[\alpha]$ in $\pi_n(B, b_0)$. Think of α as a continuous map from $(S^n, \bar{1})$ to (B, b_0). (The symbol $\bar{1}$ is used here as the base point of S^n to avoid confusion with the number 1 which will also play an important role in this proof.) Since $n \geq 2$, the fundamental group $\pi_1(S^n, \bar{1})$ is trivial, and hence

$$\alpha_* \pi_1(S^n, \bar{1}) = \{0\} \subset p_* \pi_1(E, e_0)$$

where α_* is the homomorphism induced by α on the fundamental group. Thus Theorem 5.10 shows that α has a continuous lifting

$$\tilde{\alpha}: (S^n, \bar{1}) \to (E, e_0)$$

such that $p\tilde{\alpha} = \alpha$. Then $\tilde{\alpha}$ determines a member $[\tilde{\alpha}]$ in $\pi_n(E, e_0)$ for which

$$p_*([\alpha]) = [p\tilde{\alpha}] = [\alpha],$$

so p_* maps $\pi_n(E, e_0)$ onto $\pi_n(B, b_0)$.

To see that p_* is one-to-one, suppose that $[\beta]$ belongs to its kernel; i.e.,

$$p_*([\beta]) = [p\beta] = [c]$$

where c is the constant map $c(S^n) = b_0$. As maps from $(S^n, \bar{1})$ to (B, b_0), $p\beta$ and c are equivalent, so there is a homotopy $H: S^n \times I \to B$ satisfying

$$H(t, 0) = p\beta(t), \qquad H(t, 1) = b_0, \qquad t \in S^n,$$
$$H(\bar{1}, s) = b_0, \qquad s \in I.$$

The fundamental group $\pi_1(S^n \times I, (\bar{1}, 0))$ is trivial since $n \geq 2$, so Theorem 5.10 applies again to show the existence of a lifting

$$\tilde{H}: S^n \times I \to E$$

such that

$$p\tilde{H} = H, \qquad \tilde{H}(\bar{1}, 0) = e_0.$$

The lifted homotopy $\tilde{H}$ is a homotopy between β and the constant map $d(S^n) = e_0$. To see this, observe first that

$$p\tilde{H}(\cdot, 0) = p\beta, \qquad \tilde{H}(\bar{1}, 0) = \beta(\bar{1}).$$

The Corollary to Theorem 5.2 insures that $\tilde{H}(\cdot, 0) = \beta$ since S^n is connected. The same argument shows that $\tilde{H}(\cdot, 1) = d$. It remains to be seen that $\tilde{H}(\bar{1}, s) = e_0$ for each s in I. The path

$$\tilde{H}(\bar{1}, \cdot): I \to E$$

has initial point e_0 and covers the constant path $c = H(\bar{1}, \cdot)$. Since the unique covering path of c which begins at e_0 is the constant path at e_0, then

$$\tilde{H}(\bar{1}, s) = e_0, \qquad s \in I.$$

Thus $\tilde{H}: S^n \times I \to E$ is a homotopy such that

$$\tilde{H}(\cdot, 0) = \beta, \qquad \tilde{H}(\cdot, 1) = d,$$
$$\tilde{H}(\bar{1}, s) = e_0, \qquad s \in I,$$

so $[\beta] = [d]$ is the identity element of $\pi_n(E, e_0)$. Thus the kernel of p_* contains only the identity element of $\pi_n(E, e_0)$, so p_* is one-to-one. □

Example 6.5. Consider the universal covering space $(\mathbb{R}, p)$ of the unit circle S^1. By Theorem 6.9,

$$p_*: \pi_n(\mathbb{R}) \to \pi_n(S^1)$$

is an isomorphism for $n \geq 2$. But all the homotopy groups of the contractible space $\mathbb{R}$ are trivial, so

$$\pi_n(S^1) = \{0\}, \qquad n \geq 2.$$

Example 6.6. Consider the double covering (S^n, p) over projective n-space P^n. Theorem 6.9 insures that

$$\pi_k(P^n) \cong \pi_k(S^n), \qquad k \geq 2, n \geq 2.$$

Recalling Example 6.3, we have

$$\pi_n(P^n) \cong \mathbb{Z}, \qquad n \geq 2.$$

6.4 Homotopy Equivalence

This section examines an equivalence relation for topological spaces which was introduced by Hurewicz in 1936. The relation is weaker than homeomorphism but strong enough to insure that equivalent spaces have isomorphic homotopy groups in corresponding dimensions.

Definition. Let X and Y be topological spaces. Then X and Y are *homotopy equivalent* or have the *same homotopy type* provided that there exist continuous maps $f: X \to Y$ and $g: Y \to X$ for which the composite maps gf and fg are homotopic to the identity maps on X and Y respectively. The map f is called a *homotopy equivalence*, and g is a *homotopy inverse* for f.

It should be clear that homeomorphic spaces are homotopy equivalent.

Theorem 6.10. *The relation "X is homotopy equivalent to Y" is an equivalence relation for topological spaces.*

PROOF. The relation is reflexive since the identity map on any space X is a homotopy equivalence. The symmetric property is implicit in the definition; note that both f and g are homotopy equivalences and that each is a homotopy inverse for the other.

To see that the relation is transitive, let $f: X \to Y$ and $h: Y \to Z$ be homotopy equivalences with homotopy inverses $g: Y \to X$ and $k: Z \to Y$ respectively. We must show that X and Z are homotopy equivalent. The most likely candidate for a homotopy equivalence between X and Z is hf with gk as the leading contender for homotopy inverse. Let $L: Y \times I \to Y$ be a homotopy such that $L(\cdot, 0) = kh$ and $L(\cdot, 1)$ is the identity map on Y. Then the map $M: X \times I \to X$ defined by

$$M(x, t) = gL(f(x), t), \qquad (x, t) \in X \times I,$$

is a homotopy such that

$$M(\cdot, 0) = gL(f(\cdot), 0) = (gk)(hf),$$
$$M(\cdot, 1) = gL(f(\cdot), 1) = gf,$$

so $(gk)(hf)$ is homotopic to gf and hence homotopic to the identity map on X. A completely analogous argument shows that $(hf)(gk)$ is homotopic to the identity on Z, so X and Z are homotopy equivalent. □

Example 6.7. A circle and an annulus are homotopy equivalent. To see this, consider the unit circle S^1 and the annulus $A = \{y \in \mathbb{R}^2 : 1 \leq |y| \leq 2\}$ shown in Figure 6.2.

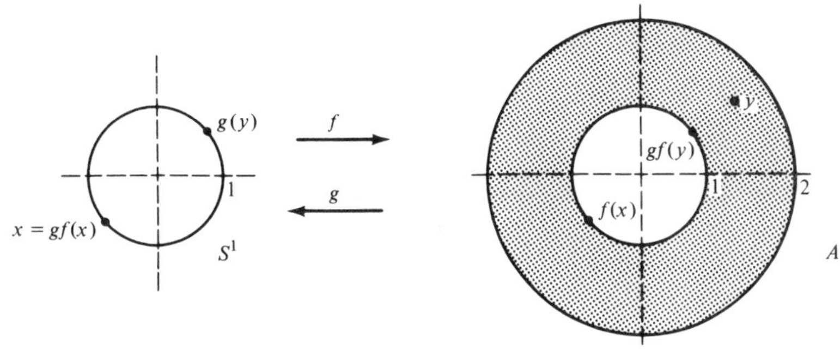

Figure 6.2

A homotopy equivalence $f: S^1 \to A$ and homotopy inverse $g: A \to S^1$ are defined by

$$f(x) = x, \qquad x \in S^1,$$
$$g(y) = y/|y|, \qquad y \in A.$$

Then gf is the identity map on S^1, and

$$fg(y) = y/|y|, \qquad y \in A.$$

The required homotopy between fg and the identity on A is given by

$$H(y, t) = ty + (1 - t)y/|y|.$$

Theorem 6.11. *A space X is contractible if and only if it has the homotopy type of a one point space.*

Proof. Suppose X is contractible with homotopy $H: X \times I \to X$ and point x_0 in X such that

$$H(x, 0) = x, \qquad H(x, 1) = x_0, \qquad x \in X.$$

Then X is homotopy equivalent to the singleton space $\{x_0\}$ by homotopy equivalence $f: X \to \{x_0\}$ and homotopy inverse $g: \{x_0\} \to X$ defined by

$$f(x) = x_0, \qquad g(x_0) = x_0, \qquad x \in X.$$

Suppose now that $f: X \to \{a\}$ is a homotopy equivalence between X and the one point space $\{a\}$ with homotopy inverse $g: \{a\} \to X$. Then there is a homotopy K between gf and the identity map on X:

$$K(x, 0) = x, \qquad K(x, 1) = gf(x) = g(a), \qquad x \in X.$$

The homotopy K is thus a contraction, and X is contractible. □

Example 6.7 is a special case of the next result.

Theorem 6.12. *If X is a space and D a deformation retract of X, then D and X are homotopy equivalent.*

PROOF. There is a homotopy $H: X \times I \to X$ such that

$$H(x, 0) = x, \qquad H(x, 1) \in D, \qquad x \in X,$$
$$H(a, t) = a, \qquad a \in D, t \in I.$$

Let $f: D \to X$ denote the inclusion map $f(a) = a$, and define $g: X \to D$ by

$$g(x) = H(x, 1), \qquad x \in X.$$

Then gf is the identity map on D, and H is a homotopy between fg and the identity on X; thus f is a homotopy equivalence with homotopy inverse g. □

Definition. Let X and Y be spaces with points x_0 in X and y_0 in Y. Then the pairs (X, x_0) and (Y, y_0) are *homotopy equivalent* or have the *same homotopy type* means that there exist continuous maps $f: (X, x_0) \to (Y, y_0)$ and $g: (Y, y_0) \to (X, x_0)$ for which the composite maps gf and fg are homotopic to the identity maps on X and Y respectively by homotopies that leave the base points fixed. In other words, it is required that there exist homotopies $H: X \times I \to X$ and $K: Y \times I \to Y$ such that

$$H(x, 0) = gf(x), \qquad H(x, 1) = x, \qquad H(x_0, t) = x_0, \qquad x \in X, t \in I,$$
$$K(y, 0) = fg(y), \qquad K(y, 1) = y, \qquad K(y_0, t) = y_0, \qquad y \in Y, t \in I.$$

The map f is called a *homotopy equivalence* with *homotopy inverse g*.

The proof of the next theorem is similar to the proof of Theorem 6.10 and is left as an exercise.

Theorem 6.13. *Homotopy equivalence between pairs is an equivalence relation.*

Theorem 6.14. *If the map $f: (X, x_0) \to (Y, y_0)$ is a homotopy equivalence between the indicated pairs, then the induced homomorphism*

$$f_*: \pi_n(X, x_0) \to \pi_n(Y, y_0)$$

is an isomorphism for each positive integer n.

PROOF. Let $g: (Y, y_0) \to (X, x_0)$ be a homotopy inverse for f and H a homotopy between gf and the identity map on X which leaves x_0 fixed. Let $[\alpha] \in \pi_n(X, x_0)$, and consider α as a function from I^n to X such that $\alpha(\partial I^n) = x_0$. Define a homotopy $K: I^n \times I \to X$ by

$$K(t, s) = H(\alpha(t), s), \qquad t \in I^n, s \in I.$$

Then

$$K(\cdot, 0) = gf\alpha, \qquad K(\cdot, 1) = \alpha,$$
$$K(\partial I^n \times I) = H(\{x_0\} \times I) = x_0$$

so that

$$[gf\alpha] = [\alpha].$$

This means that

$$g_* f_*[\alpha] = [\alpha],$$

so g_* is a left inverse for f_*. Since f is a homotopy inverse for g, we conclude by symmetry that g_* is also a right inverse for f_*, so f_* is an isomorphism. □

Actually, Theorem 6.14 can be strengthened to show that a homotopy equivalence $f: X \to Y$ with $f(x_0) = y_0$ induces an isomorphism between $\pi_n(X, x_0)$ and $\pi_n(Y, y_0)$ for each n. The proof is more complicated because the homotopies may not leave the base points fixed. The reader might like to try proving this stronger result.

6.5 Homotopy Groups of Spheres

As mentioned earlier, the homotopy groups $\pi_k(S^n)$ are not completely known. Previous examples have shown that

$$\begin{aligned} \pi_k(S^n) &= \{0\}, \qquad k < n, \\ \pi_k(S^1) &= \{0\}, \qquad k > 1, \\ \pi_n(S^n) &\cong \mathbb{Z}. \end{aligned}$$

It may seem natural to conjecture that $\pi_k(S^n)$ is trivial for $k > n$ since the corresponding result holds for the homology groups. This would simply mean that every continuous map $f: S^k \to S^n$ where $k > n$ is homotopic to a constant map. This is in fact not true. The first example of such an essential, or non-null-homotopic, map was given by H. Hopf in 1931. The spheres involved were of dimensions three and two, and Hopf's example showed that $\pi_3(S^2)$ is not trivial. Actually, $\pi_3(S^2)$ is isomorphic to the group of integers. Many other results are known about $\pi_k(S^n)$, but no one has yet succeeded in determining $\pi_k(S^n)$ in all possible cases. In this section we shall examine Hopf's examples and the results of H. Freudenthal (1905–) on which much of the knowledge of the higher homotopy groups of spheres is based.

Example 6.8. The Hopf map $p: S^3 \to S^2$.

Let $\mathbb{C}$ denote the field of complex numbers. Consider S^3, the unit sphere in Euclidean 4-space, as a set of ordered pairs of complex numbers, each pair having length 1:

$$S^3 = \{(z_1, z_2) \in \mathbb{C} \times \mathbb{C}: |z_1|^2 + |z_2|^2 = 1\}.$$

Define an equivalence relation $\equiv$ on S^3 by

$$(z_1, z_2) \equiv (z_1', z_2')$$

if and only if there is a complex number λ of length 1 such that

$$(z_1, z_2) = (\lambda z_1', \lambda z_2').$$

For (z_1, z_2) in S^3, let $\langle z_1, z_2 \rangle$ denote the equivalence class determined by (z_1, z_2), let

$$T = \{\langle z_1, z_2 \rangle: (z_1, z_2) \in S^3\}$$

be the set of equivalence classes, and let $p: S^3 \to T$ be the projection map

$$p(z_1, z_2) = \langle z_1, z_2 \rangle, \qquad (z_1, z_2) \in S^3.$$

Assign T the quotient topology determined by p; a set O is open in T provided that $p^{-1}(O)$ is open in S^3. For $\langle z_1, z_2 \rangle$ in T, the inverse image $p^{-1}(\langle z_1, z_2 \rangle)$, called the *fiber* over $\langle z_1, z_2 \rangle$, is a circle in S^3.

We shall show that T is homeomorphic to S^2, use the homeomorphism to replace T by S^2, and obtain the Hopf map $p: S^3 \to S^2$. Strictly speaking, the Hopf map is the map $hp: S^3 \to S^2$ where $h: T \to S^2$ is the homeomorphism whose existence we must now show.

Let

$$D = \{z \in \mathbb{C} : |z| \leq 1\}$$

denote the unit disc in $\mathbb{C}$. The 2-sphere is the quotient space of D obtained by identifying the boundary of D to a point. To see that T satisfies the same description, consider the map $f: D \to T$ defined by

$$f(z) = \langle \sqrt{1 - |z|^2}, z \rangle, \qquad z \in D.$$

Then f is a closed, continuous map. For $\langle z_1, z_2 \rangle$ in T,

$$\begin{aligned} f^{-1}(\langle z_1, z_2 \rangle) &= \{z \in D : \langle z_1, z_2 \rangle = \langle \sqrt{1 - |z|^2}, z \rangle\} \\ &= \{z \in D : \sqrt{1 - |z|^2} = \lambda z_1, z = \lambda z_2 \text{ for some } \lambda \in S^1\}. \end{aligned}$$

If $z_1 \neq 0$, the equations

$$\sqrt{1 - |z|^2} = \lambda z_1, \qquad z = \lambda z_2, \qquad |z_1|^2 + |z_2|^2 = 1$$

imply

$$\lambda z_1 = |z_1|, \qquad \lambda = |z_1|/z_1.$$

Thus $f^{-1}(\langle z_1, z_2 \rangle)$ is a single point if $z_1 \neq 0$. If $z_1 = 0$, then

$$\begin{aligned} f^{-1}(\langle z_1, z_2 \rangle) &= f^{-1}(\langle 0, z_2 \rangle) \\ &= \{z \in D : \sqrt{1 - |z|^2} = 0, z = \lambda \text{ for some } \lambda \in S^1\} = S^1, \end{aligned}$$

so $f^{-1}(\langle 0, z_2 \rangle)$ is the boundary of D. Hence, using f as quotient map, T is the quotient space of D obtained by identifying the boundary S^1 to a point. Then T is homeomorphic to S^2, so we replace T by S^2 and have the Hopf map $p: S^3 \to S^2$.

Showing that p is not homotopic to a constant map requires more background than we have had, but here is a sketch of the basic idea. Suppose to the contrary that $H: S^3 \times I \to S^2$ is a homotopy between p and a constant map. Although the Hopf map is not a covering projection, it is close enough to permit a covering homotopy $\tilde{H}: S^3 \times I \to S^2$ as shown in this diagram.

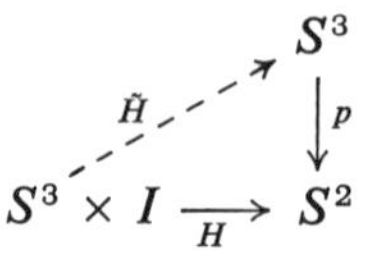

The map $\tilde{H}$ is a homotopy between the identity map on S^3 and a constant map. But this implies that S^3 is contractible, an obvious contradiction. Thus p is not homotopic to a constant map, so $\pi_3(S^2) \neq \{0\}$.

Example 6.9. The Hopf maps $S^7 \to S^4$ and $S^{15} \to S^8$.

Think for a minute about the construction of the Hopf map $p: S^3 \to S^2$. The construction was made possible by representing S^3 as ordered pairs of complex numbers. Using the division ring $\mathbb{Q}$ of quaternions, we represent S^7, the unit sphere in Euclidean 8-space, as ordered pairs of members of $\mathbb{Q}$:

$$S^7 = \{(z_1, z_2) \in \mathbb{Q}: \|z_1\|^2 + \|z_2\|^2 = 1\}.$$

The quotient space T in this case is the quotient space of the unit disc

$$D = \{z \in \mathbb{Q}: \|z\| \leq 1\}$$

obtained by identifying the boundary of D to a single point. Since D has real dimension four, this quotient space is homeomorphic to S^4. The Hopf map $p: S^7 \to S^4$ with fiber S^3 is then defined as in Example 6.8. This map shows that $\pi_7(S^4) \neq \{0\}$.

In E^{16}, one can perform a similar construction by representing the unit sphere S^{15} as ordered pairs of Cayley numbers. This produces the Hopf map $p: S^{15} \to S^8$ with fiber S^7 and shows that $\pi_{15}(S^8) \neq \{0\}$.

There is for each pair k, n of positive integers a natural homomorphism

$$E: \pi_k(S^n) \to \pi_{k+1}(S^{n+1})$$

called the *suspension homomorphism*. To define this ingenious function, consider $\pi_k(S^n)$ as homotopy classes of maps from $(S^k, 1)$ to $(S^n, 1)$ where we denote the base point of each sphere by 1. Consider S^n as the subspace of S^{n+1} consisting of all points of S^{n+1} having last coordinate 0. In this identification, S^n is usually called the "equator" of S^{n+1}. Continuing this geographical metaphor, call the points $(0, \ldots, 0, 1)$ and $(0, \ldots, 0, -1)$ of S^{n+1} the "north pole" and "south pole" respectively.

Suppose now that $[\alpha] \in \pi_k(S^n)$. Then α is a continuous map from S^k to S^n. Extend α to a continuous map $\hat{\alpha}: S^{k+1} \to S^{n+1}$ as follows: $\hat{\alpha}|_{S^k}$ is just α, and it maps the equator of S^{k+1} to the equator of S^{n+1}. We require that $\hat{\alpha}$ map the north pole of S^{k+1} to the north pole of S^{n+1} and the south pole of S^{k+1} to the south pole of S^{n+1}. The function is then extended radially as shown in Figure 6.3. The arc from the north pole to a point x in S^k is mapped linearly onto the arc from the north pole of S^{n+1} to $\alpha(x)$. This defines $\hat{\alpha}$ on the "northern hemisphere," and the "southern hemisphere" is treated the same way. The extended map $\hat{\alpha}$ is called the *suspension* of α.

The *suspension homomorphism* E, called the "Einhängung" by Freudenthal, is defined by

$$E([\alpha]) = [\hat{\alpha}], \qquad [\alpha] \in \pi_k(S^n).$$

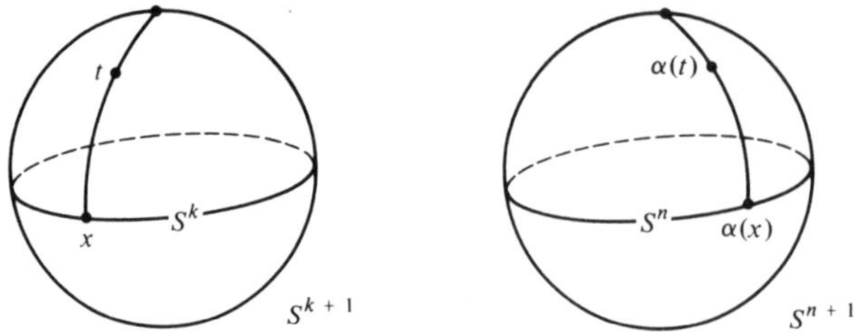

Figure 6.3

The reader is left the exercise of showing that E is a homomorphism. Freudenthal defined the suspension homomorphism and proved the following theorem in 1937. Proofs can be found in [11] and in Freudenthal's original paper [37].

Theorem 6.15 (The Freudenthal Suspension Theorem). *The suspension homomorphism*

$$E: \pi_k(S^n) \to \pi_{k+1}(S^{n+1})$$

is an isomorphism for $k < 2n - 1$ *and is onto for* $k \leq 2n - 1$.

Although we shall not prove the Freudenthal Suspension Theorem, we illustrate its utility with two corollaries. These results have already been derived in Examples 6.2 and 6.3.

Corollary. *The homotopy groups* $\pi_k(S^n)$ *are trivial for* $k < n$.

PROOF. For any positive integer $r < k$, we have $k + r + 1 < 2n$, and hence

$$k - r < 2(n - r) - 1.$$

Then

$$\pi_k(S^n) \cong \pi_{k-1}(S^{n-1}) \cong \cdots \cong \pi_1(S^{n-k+1}).$$

Since $n - k + 1 > 1$ for $k < n$, then $\pi_1(S^{n-k+1})$ and its isomorphic image $\pi_k(S^n)$ are both trivial groups. □

Corollary. *The homotopy groups* $\pi_n(S^n)$, $n \geq 1$, *are all isomorphic to the group* $\mathbb{Z}$ *of integers.*

PROOF. We rely on our previous arguments to show that

$$\pi_1(S^1) \cong \pi_2(S^2) \cong \mathbb{Z}.$$

If $n \geq 2$, then $n < 2n - 1$ and the Freudenthal Suspension Theorem shows that

$$\pi_2(S^2) \cong \pi_3(S^3) \cong \pi_4(S^4) \cong \cdots \cong \pi_n(S^n).$$ □

6.6 The Relation Between $H_n(K)$ and $\pi_n(|K|)$

The last theorem of this chapter extends Theorem 4.11 to show a relationship between the homology groups and the homotopy groups of polyhedra. Proofs can be found in [20] and [5]

Theorem 6.16 (The Hurewicz Isomorphism Theorem). *Let K be a connected complex and $n \geq 2$ a positive integer. If the first $n - 1$ homotopy groups of $|K|$ are trivial, then $H_n(K)$ and $\pi_n(|K|)$ are isomorphic.*

For an application of the Hurewicz Isomorphism Theorem, let us again consider $\pi_n(S^n)$.

Example 6.10. Consider the n-sphere S^n for $n \geq 2$. Since $\pi_k(S^n) = \{0\}$ for $k < n$, the Hurewicz Isomorphism Theorem implies that

$$\pi_n(S^n) \cong H_n(S^n) \cong \mathbb{Z}.$$

The pioneering work on the higher homotopy groups was done by Witold Hurewicz in a sequence of four papers, his famous "Four Notes," published in 1935–1936 [42]. These papers contain definitions of the higher homotopy groups, the relation between $\pi_n(E)$ and $\pi_n(B)$ for covering spaces (Theorem 6.9), the homotopy equivalence relation, the proof that homotopy equivalent spaces have isomorphic homotopy groups (Theorem 6.14), and the Hurewicz Isomorphism Theorem (Theorem 6.16).

The homotopy groups do not provide for general topological spaces the type of classification given for 2-manifolds by Theorem 2.11 and for covering spaces by Theorem 5.9. The reader is asked in one of the exercises for this chapter to find an example of spaces X and Y which have isomorphic homotopy groups in each dimension but which are not homotopy equivalent (and therefore not homeomorphic). The induced homomorphism

$$f_*: \pi_n(X) \rightarrow \pi_n(Y)$$

has been successful in classifying the homotopy type of spaces known as "CW-complexes." These spaces can be used to approximate arbitrary topological spaces. The reader interested in pursuing CW-complexes should consult [20] or the work of their inventor, J. H. C. Whitehead (1904–1960) [57].

Although the homotopy groups have not been completely successful in showing when spaces are homeomorphic, they are extremely useful in showing when spaces are not homeomorphic. This is, in fact, the way in which algebraic topology has been most successful. To show that X and Y are not homeomorphic, compute the homotopy groups $\pi_n(X)$ and $\pi_n(Y)$. If $\pi_n(X)$ is not isomorphic to $\pi_n(Y)$ for some n, then X and Y are not homeomorphic. The same method can be used with the homology groups.

Recall from Chapter 4 that the Poincaré Conjecture asserts that every simply connected 3-manifold is homeomorphic to S^3. Our work on homotopy groups shows that the corresponding conjecture in dimension four is false. The 4-manifold $S^2 \times S^2$ is simply connected, but it is not homeomorphic to S^4 since

$$\pi_2(S^2 \times S^2) \cong \mathbb{Z} \oplus \mathbb{Z} \quad \text{and} \quad \pi_2(S^4) = \{0\}.$$

Hurewicz' introduction of homotopy type led to the following extension of the Poincaré Conjecture:

Generalized Poincaré Conjecture. *Every n-manifold which is homotopy equivalent to S^n is homeomorphic to S^n.*

This conjecture was proved to be true for $n > 4$ by S. Smale (1930–) in 1960 [54]. It is still unresolved in the cases $n = 3$ and $n = 4$.

EXERCISES

1. Complete the details in Definition A of the higher homotopy groups:
   (a) The relation $\sim_{x_0}$ is an equivalence relation.
   (b) If $\alpha \sim_{x_0} \alpha'$ and $\beta \sim_{x_0} \beta'$, then $\alpha * \beta \sim_{x_0} \alpha' * \beta'$.
   (c) $\pi_n(X, x_0)$ is a group under the operation $\circ$.
2. Complete the details in the discussion of the equivalence of Definitions A and C of the higher homotopy groups.
3. (a) Prove Theorem 6.2.
   (b) Prove Theorem 6.3.
   (c) Prove Theorem 6.4.
4. Let $f: X \to S^n$ be a continuous map such that $f(X)$ is a proper subset of S^n. Prove that f is null-homotopic.
5. Use homotopy groups to prove the Brouwer No Retraction Theorem (Theorem 3.12).
6. Show that the sets $W(K, U)$ in the definition of the compact-open topology form a subbase.
7. (a) Show that the space $\Omega(X, x_0)$ with its compact-open topology is an H-space for any space X.
   (b) Show that the homotopy classes $[\alpha]$ of $\pi_1(X, x_0)$ are precisely the path components of $\Omega(X, x_0)$.
8. Show that the function $f_*: \pi_n(X, x_0) \to \pi_n(Y, y_0)$ induced by a continuous map $f: (X, x_0) \to (Y, y_0)$ is a homomorphism.
9. Prove that the operation $\square$ in Theorems 6.5 and 6.6 is well-defined.
10. If $f: X \to Y$ is a homotopy equivalence, prove that any two homotopy inverses of f are homotopic.
11. **Definition.** If $f: X \to Y$ is a continuous map, a continuous map $g: Y \to X$ is a *left homotopy inverse* for f provided that gf is homotopic to the identity map on X. *Right homotopy inverse* is defined analogously.

    Prove that if $f: X \to Y$ has left homotopy inverse g and right homotopy inverse h, then f is a homotopy equivalence.
12. **Definition.** Continuous maps f and g from (X, x_0) to (Y, y_0) are *homotopic modulo base points* provided that there is a homotopy $H: X \times I \to Y$ such that

$$H(\cdot, 0) = f, \qquad H(\cdot, 1) = g, \qquad H(\{x_0\} \times I) = y_0.$$

    Prove that maps which are homotopic modulo base points induce identical homomorphisms from $\pi_n(X, x_0)$ to $\pi_n(Y, y_0)$.

13. Prove that the map f of Example 6.8 is closed and continuous.

14. If X is homotopy equivalent to X' and Y is homotopy equivalent to Y', prove that $X \times Y$ is homotopy equivalent to $X' \times Y'$.

15. Show that if the pairs (X, x_0) and (Y, y_0) are homotopy equivalent, then the loop spaces $\Omega(X, x_0)$ and $\Omega(Y, y_0)$ are homotopy equivalent.

16. Let (E, p) and (F, q) be covering spaces of base space B, and let $h: E \to F$ be a covering space homomorphism such that $h(e_0) = f_0$, where e_0 and f_0 are the base points of E and F respectively. Show that the induced homomorphism

$$h_*: \pi_n(E, e_0) \to \pi_n(F, f_0)$$

is an isomorphism for $n \geq 2$. What can be said about h_* if $n = 1$?

17. Show that the Freudenthal map

$$E: \pi_k(S^n) \to \pi_{k+1}(S^{n+1})$$

is a homomorphism.

18. **Definition.** Let $f: X \to Y$ be a continuous map. The quotient space of the disjoint union $(X \times I) \cup Y$ obtained by identifying $(x, 1)$ with $f(x)$, $x \in X$, is called the *mapping cylinder* of f.

Show that the mapping cylinder of $f: X \to Y$ is homotopy equivalent to Y.

19. Show that the unit sphere S^{n-1} and punctured n-space $\mathbb{R}^n \backslash \{p\}$ have the same homotopy type.

20. Here are some homotopy groups of spheres. Use them to determine other homotopy groups of spheres. (The symbol $\mathbb{Z}_p$ denotes the group of integers modulo p).

(a) $\pi_{12}(S^7) = \{0\}$.
(b) $\pi_{14}(S^8) \cong \mathbb{Z}$.
(c) $\pi_{16}(S^9) \cong \mathbb{Z}_{240}$.
(d) $\pi_{18}(S^{10}) \cong \mathbb{Z}_2 \oplus \mathbb{Z}_2$.

21. Prove that homotopy equivalence for pairs is an equivalence relation.

22. Give an example of spaces X and Y having isomorphic homotopy groups in each dimension which do not have the same homotopy type.

7 Further Developments in Homology

The preceding chapters have introduced homology groups for polyhedra and homotopy groups for arbitrary spaces. The homotopy groups are more general since they apply to more spaces. The process of extending homology to spaces more general than polyhedra began in the years 1921–1933 and has continued to the present day. The pioneers in this work were Oswald Veblen, Solomon Lefschetz, Leopold Vietoris, and Eduard Čech. In this chapter we shall examine some additional theory and applications of simplicial homology groups, notably the famous fixed point theorem and relative homology groups discovered by Lefschetz, and the singular homology groups, also due to Lefschetz, which extend homology theory to arbitrary spaces.

7.1 Chain Derivation

Chain mappings were introduced in Chapter 3 for the purpose of defining induced homomorphisms on the homology groups. We turn now to a particular chain mapping, the "chain derivation" $\varphi = \{\varphi_p: C_p(K) \to C_p(K^{(1)})\}$, from the chain groups of a complex K to those of its first barycentric subdivision $K^{(1)}$. This will allow us to see that $H_p(K) \cong H_p(K^{(1)})$, a problem that was glossed over in Chapter 3, and to establish the machinery necessary for a proof of Lefschetz' celebrated fixed point theorem.

Notation: If $\sigma^p = \langle v_0 \ldots v_p \rangle$ is a p-simplex and v a vertex for which $\{v, v_0, \ldots, v_p\}$ is geometrically independent, then the symbol $v\sigma^p$ denotes the $(p + 1)$-simplex

$$v\sigma^p = \langle vv_0 \ldots v_p \rangle.$$

If $c = \sum g_i \cdot \sigma_i^p$ is a p-chain, then vc denotes the $(p + 1)$-chain

$$vc = \sum g_i \cdot v\sigma_i^p$$

This notation was used in Theorem 2.9.

The proof of the following lemma is left as an exercise:

Lemma. *Let c be a p-chain on a complex K and v a vertex for which the $(p+1)$-chain vc is defined. Then*

$$\partial(vc) = c - v\partial c.$$

Definition. Let K be a complex. A chain mapping

$$\varphi = \{\varphi_p : C_p(K) \to C_p(K^{(1)})\}$$

is defined inductively as follows: Each 0-simplex σ^0 of K is a 0-simplex of the barycentric subdivision $K^{(1)}$, so we may consider $C_0(K)$ as a subgroup of $C_0(K^{(1)})$. Define $\varphi_0 : C_0(K) \to C_0(K^{(1)})$ to be the inclusion map:

$$\varphi_0(c) = c, \qquad c \in C_0(K).$$

For an elementary p-chain $1 \cdot \sigma^p$ on K, define

$$\varphi_p(1 \cdot \sigma^p) = \dot{\sigma}^p \varphi_{p-1}\partial(1 \cdot \sigma^p),$$

where $\dot{\sigma}^p$ denotes the barycenter of σ^p, and extend φ_p by linearity to a homomorphism $\varphi_p : C_p(K) \to C_p(K^{(1)})$:

$$\varphi_p\left(\sum g_i \cdot \sigma_i^p\right) = \sum \varphi_p(g_i \cdot \sigma_i^p), \qquad \sum g_i \cdot \sigma_i^p \in C_p(K).$$

The sequence $\varphi = \{\varphi_p\}$ of homomorphisms defined in this way is the *first chain derivation* on K. For $n > 1$, the nth *chain derivation* on K is the composition of $\varphi^{(n-1)}$, the $(n-1)$th chain derivation on K, with the first chain derivation of the $(n-1)$th barycentric subdivision $K^{(n-1)}$. Thus the nth chain derivation on K is a chain mapping $\varphi^{(n)} = \{\varphi_p^{(n)} : C_p(K) \to C_p(K^{(n)})\}$.

Example 7.1. Let us examine the first chain derivation of the complex $K = \mathrm{Cl}(\sigma^2)$, the closure of a 2-simplex $\sigma^2 = +\langle v_0v_1v_2\rangle$, shown with the barycentric subdivision $K^{(1)}$ in Figure 7.1.

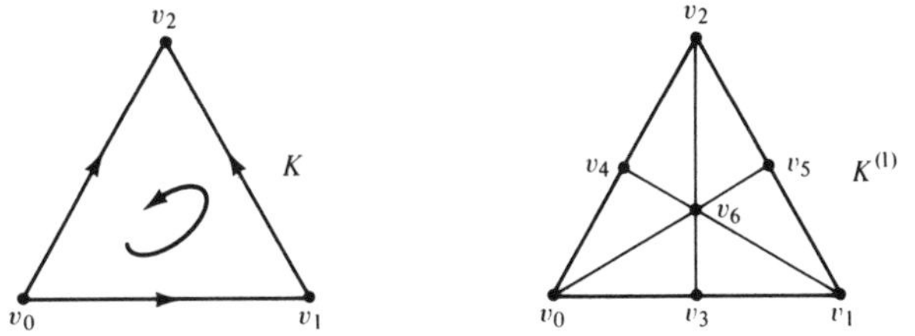

Figure 7.1

In the figure, the additional vertices v_3, v_4, v_5, and v_6 denote the barycenters of $\langle v_0v_1\rangle$, $\langle v_0v_2\rangle$, $\langle v_1v_2\rangle$, and $\langle v_0v_1v_2\rangle$ respectively. Then $\varphi_0 : C_0(K) \to C_0(K^{(1)})$ is the inclusion map, and

$$\begin{aligned}\varphi_1(1 \cdot \langle v_0v_1\rangle) &= v_3\varphi_0\partial(1 \cdot \langle v_0v_1\rangle) = v_3(1 \cdot \langle v_1\rangle - 1 \cdot \langle v_0\rangle)\\ &= 1 \cdot \langle v_3v_1\rangle - 1 \cdot \langle v_3v_0\rangle;\end{aligned}$$

$$\begin{aligned}
\varphi_1(1\cdot\langle v_0v_2\rangle) &= v_4\varphi_0\partial(1\cdot\langle v_0v_2\rangle) = v_4(1\cdot\langle v_2\rangle - 1\cdot\langle v_0\rangle)\\
&= 1\cdot\langle v_4v_2\rangle - 1\cdot\langle v_4v_0\rangle;\\
\varphi_1(1\cdot\langle v_1v_2\rangle) &= v_5\varphi_0\partial(1\cdot\langle v_1v_2\rangle) = v_5(1\cdot\langle v_2\rangle - 1\cdot\langle v_1\rangle)\\
&= 1\cdot\langle v_5v_2\rangle - 1\cdot\langle v_5v_1\rangle;\\
\varphi_2(1\cdot\langle v_0v_1v_2\rangle) &= v_6\varphi_1\partial(1\cdot\langle v_0v_1v_2\rangle) = v_6\varphi_1(1\cdot\langle v_1v_2\rangle - 1\cdot\langle v_0v_2\rangle + 1\cdot\langle v_0v_1\rangle)\\
&= 1\cdot\langle v_6v_5v_2\rangle - 1\cdot\langle v_6v_5v_1\rangle - 1\cdot\langle v_6v_4v_2\rangle + 1\cdot\langle v_6v_4v_0\rangle\\
&\quad + 1\cdot\langle v_6v_3v_1\rangle - 1\cdot\langle v_6v_3v_0\rangle.
\end{aligned}$$

Theorem 7.1. *Each chain derivation is a chain mapping.*

Proof. Since the composition of chain mappings is a chain mapping, it is sufficient to show that the first chain derivation is a chain mapping. Let $\varphi = \{\varphi_p: C_p(K) \to C_p(K^{(1)})\}$ be a chain derivation in the notation of the definition. It must be shown that the diagram

$$\begin{array}{ccc}
C_p(K) & \xrightarrow{\varphi_p} & C_p(K^{(1)})\\
\partial\downarrow & & \downarrow\partial\\
C_{p-1}(K) & \xrightarrow{\varphi_{p-1}} & C_{p-1}(K^{(1)})
\end{array}$$

is commutative for $p \geq 1$. Thus it is sufficient to show that

$$\partial\varphi_p(1\cdot\sigma^p) = \varphi_{p-1}\partial(1\cdot\sigma^p)$$

for each elementary p-chain $1\cdot\sigma^p$. For $p = 1$,

$$\begin{aligned}
\partial\varphi_1(1\cdot\sigma^1) &= \partial(\dot\sigma^1\varphi_0\partial(1\cdot\sigma^1)) = \varphi_0\partial(1\cdot\sigma^1) - \dot\sigma^1\partial\varphi_0\partial(1\cdot\sigma^1)\\
&= \varphi_0\partial(1\cdot\sigma^1) - \dot\sigma^1\partial\partial(1\cdot\sigma^1) = \varphi_0\partial(1\cdot\sigma^1).
\end{aligned}$$

These equalities follow, in order, from the definition of φ_1, the lemma $\partial(vc) = c - v\partial c$, the fact that φ_0 is the inclusion map, and $\partial\partial = 0$. Thus $\partial\varphi_1 = \varphi_0\partial$, so the desired conclusion holds for $p = 1$. Proceeding inductively, let $1\cdot\sigma^p$ be an elementary p-chain on K. Then

$$\begin{aligned}
\partial\varphi_p(1\cdot\sigma^p) &= \partial(\dot\sigma^p\varphi_{p-1}\partial(1\cdot\sigma^p)) = \varphi_{p-1}\partial(1\cdot\sigma^p) - \dot\sigma^p\partial\varphi_{p-1}\partial(1\cdot\sigma^p)\\
&= \varphi_{p-1}\partial(1\cdot\sigma^p) - \dot\sigma^p\varphi_{p-2}\partial\partial(1\cdot\sigma^p) = \varphi_{p-1}\partial(1\cdot\sigma^p).
\end{aligned}$$

The next to last equality uses the inductive assumption $\partial\varphi_{p-1} = \varphi_{p-2}\partial$. Thus $\partial\varphi_p = \varphi_{p-1}\partial$ for elementary p-chains and hence for all p-chains. □

Theorem 7.2. *Let K be a complex with first chain derivation $\varphi = \{\varphi_p\}$. There is a chain mapping*

$$\psi = \{\psi_p: C_p(K^{(1)}) \to C_p(K)\}$$

such that $\psi_p\varphi_p$ is the identity map on $C_p(K)$ for each $p \geq 0$.

Proof. Such a chain mapping ψ is called a *left inverse* for φ. Let f be any simplicial map from $K^{(1)}$ to K having this property: If $\dot\sigma$ is a vertex of $K^{(1)}$, then $f(\dot\sigma)$ is a vertex of the simplex σ of which $\dot\sigma$ is the barycenter. Let $\psi = \{\psi_p\}$ be the chain mapping induced by f. Observe that if τ^p is a p-simplex of $K^{(1)}$, then

$$\psi_p(1\cdot\tau^p) = \eta\cdot\sigma^p,$$

where η is 0, 1 or -1 and σ^p is the p-simplex of K which produces τ^p in its barycentric subdivision.

Clearly $\psi_0\varphi_0$ is the identity map on $C_0(K)$. Suppose that $\psi_{p-1}\varphi_{p-1}$: $C_{p-1}(K)\to C_{p-1}(K)$ is the identity, and consider $\psi_p\varphi_p$: $C_p(K)\to C_p(K)$. If $1\cdot\sigma^p$ is an elementary p-chain on K, then

$$\psi_p\varphi_p(1\cdot\sigma^p) = \psi_p(\dot{\sigma}^p\varphi_{p-1}\partial(1\cdot\sigma^p)) = m\cdot\sigma^p$$

for some integer m. But

$$\partial(m\cdot\sigma^p) = \partial\psi_p\varphi_p(1\cdot\sigma^p) = \psi_{p-1}\partial\varphi_p(1\cdot\sigma^p) = \psi_{p-1}\varphi_{p-1}\partial(1\cdot\sigma^p) = \partial(1\cdot\sigma^p),$$

so

$$m\partial(1\cdot\sigma^p) = \partial(m\cdot\sigma^p) = \partial(1\cdot\sigma^p),$$

and hence $m = 1$. Thus

$$\psi_p\varphi_p(1\cdot\sigma^p) = 1\cdot\sigma^p,$$

so $\psi_p\varphi_p$ is the identity map on $C_p(K)$. □

Example 7.2. The preceding theorem is not as complicated as it may appear. Consider the chain derivation $\varphi = \{\varphi_p\}_0^2$ of Example 7.1. We may define the simplicial map f from $K^{(1)}$ to K, the closure of the 2-simplex $\langle v_0v_1v_2\rangle$, in any manner consistent with having $f(v_i)$ a vertex of the simplex of which v_i is the barycenter. Thus we must have

$$f(v_0) = v_0, \qquad f(v_1) = v_1, \qquad f(v_2) = v_2.$$

One possible definition for f on the remaining vertices is

$$f(v_3) = f(v_4) = v_0, \qquad f(v_5) = v_1, \qquad f(v_6) = v_2.$$

Let $\psi = \{\psi_p\}_0^2$ be the chain mapping induced by f, as in the proof of Theorem 7.2. Then

$$\begin{aligned}
&\psi_0(1\cdot\langle v_0\rangle) = \psi_0(1\cdot\langle v_3\rangle) = \psi_0(1\cdot\langle v_4\rangle) = 1\cdot\langle v_0\rangle;\\
&\psi_0(1\cdot\langle v_1\rangle) = \psi_0(1\cdot\langle v_5\rangle) = 1\cdot\langle v_1\rangle;\\
&\psi_0(1\cdot\langle v_2\rangle) = \psi_0(1\cdot\langle v_6\rangle) = 1\cdot\langle v_2\rangle.\\
&\psi_1(1\cdot\langle v_0v_4\rangle) = 0; \qquad \psi_1(1\cdot\langle v_0v_6\rangle) = 1\cdot\langle v_0v_2\rangle; \quad \text{etc.}\\
&\psi_2(1\cdot\langle v_3v_1v_6\rangle) = 1\cdot\langle v_0v_1v_2\rangle; \qquad \psi_2(1\cdot\langle v_0v_4v_6\rangle) = 0; \quad \text{etc.}
\end{aligned}$$

Consider, for example,

$$\psi_1\varphi_1(1\cdot\langle v_0v_1\rangle) = \psi_1(1\cdot\langle v_3v_1\rangle - 1\cdot\langle v_3v_0\rangle) = 1\cdot\langle v_0v_1\rangle - 0 = 1\cdot\langle v_0v_1\rangle.$$

Let us compute $\psi_2\varphi_2(1\cdot\langle v_0v_1v_2\rangle)$, where $\varphi_2(1\cdot\langle v_0v_1v_2\rangle)$ is expressed as in Example 7.1:

$$\begin{aligned}
\varphi_2(1\cdot\langle v_0v_1v_2\rangle) = {} & 1\cdot\langle v_6v_5v_2\rangle - 1\cdot\langle v_6v_5v_1\rangle - 1\cdot\langle v_6v_4v_2\rangle\\
& + 1\cdot\langle v_6v_4v_0\rangle + 1\cdot\langle v_6v_3v_1\rangle - 1\cdot\langle v_6v_3v_0\rangle.
\end{aligned}$$

Since f collapses all 2-simplexes except $\langle v_6v_3v_1\rangle$, then

$$\psi_2\varphi_2(1\cdot\langle v_0v_1v_2\rangle) = \psi_2(1\cdot\langle v_6v_3v_1\rangle) = 1\cdot\langle v_2v_0v_1\rangle = 1\cdot\langle v_0v_1v_2\rangle.$$

Definition. A pair $\varphi = \{\varphi_p\}_0^\infty$ and $\mu = \{\mu_p\}_0^\infty$ of chain mappings from a complex K to a complex L are *chain homotopic* means that there is a sequence $\mathscr{D} = \{D_p\}_{-1}^\infty$ of homomorphisms $D_p\colon C_p(K) \to C_{p+1}(L)$ such that

$$\partial D_p + D_{p-1}\partial = \varphi_p - \mu_p, \qquad D_{-1} = 0.$$

The sequence $\mathscr{D}$ is called a *deformation operator* or a *chain homotopy.*

The chain homotopy relation was designed explicitly to produce the next theorem.

Theorem 7.3. *If φ and μ are chain homotopic chain mappings from complex K to complex L, then the induced homomorphisms φ_p^* and μ_p^* from $H_p(K)$ to $H_p(L)$ are equal, $p \geq 0$.*

Proof. Since φ and μ are chain homotopic, there is a deformation operator $\mathscr{D} = \{D_p\}_{-1}^\infty$ as specified in the definition. For $[z_p] \in H_p(K)$,

$$\varphi_p^*([z_p]) - \mu_p^*([z_p]) = [\varphi_p(z_p) - \mu_p(z_p)] = [\partial D_p(z_p) + D_{p-1}(\partial z_p)] = 0.$$

The final equality follows because $\partial z_p = 0$ for any cycle and $\partial D_p(z_p)$ is a boundary. Thus $\varphi_p^* = \mu_p^*$ for each value of p. □

Definition. Complexes K and L are *chain equivalent* means that there are chain mappings φ from K to L and ψ from L to K such that the composite chain mappings $\psi\varphi = \{\psi_p\varphi_p\}$ and $\varphi\psi = \{\varphi_p\psi_p\}$ are chain homotopic to the identity chain mappings on K and L respectively.

It is left to the reader to show that chain homotopy is an equivalence relation for chain mappings and that chain equivalence is an equivalence relation for complexes.

Theorem 7.4. *Chain equivalent complexes K and L have isomorphic homology groups in corresponding dimensions.*

Proof. If φ and ψ are the chain mappings required by the definition of chain equivalence, then Theorem 7.3 insures that

$$\psi_p^*\varphi_p^*\colon H_p(K) \to H_p(K),$$
$$\varphi_p^*\psi_p^*\colon H_p(L) \to H_p(L)$$

are the identity maps, so φ_p^* is an isomorphism for each value of p. □

One objective of this section is to prove that the homology groups of a complex K are isomorphic to those of its barycentric subdivision $K^{(1)}$. In view of Theorem 7.4, it is sufficient to show that K and $K^{(1)}$ are chain equivalent.

For this we need chain mappings φ from K to $K^{(1)}$ and ψ from $K^{(1)}$ to K for which $\psi\varphi$ and $\varphi\psi$ are chain homotopic to the appropriate identity chain maps. We have φ, the first chain derivation of K; we also have ψ, the left inverse provided by Theorem 7.2. We know that $\psi\varphi$ is the identity chain map on K, and we must show that $\varphi\psi$ is chain homotopic to the identity chain map on $K^{(1)}$. This is a rather large assignment; it is accomplished by the next proof.

Theorem 7.5. *A complex K and its first barycentric subdivision are chain equivalent.*

PROOF. In view of the preceding discussion, it is sufficient to show that $\varphi\psi$ is chain homotopic to the identity map on $K^{(1)}$. This requires a deformation operator $\mathscr{D} = \{D_p: C_p(K^{(1)}) \to C_{p+1}(K^{(1)})\}$ such that $D_{-1} = 0$ and, for each elementary p-chain $1 \cdot \tau^p$ on $K^{(1)}$,

$$1 \cdot \tau^p - \varphi_p\psi_p(1 \cdot \tau^p) = \partial D_p(1 \cdot \tau^p) - D_{p-1}\partial(1 \cdot \tau^p).$$

We must have $D_{-1} = 0$. To define D_0, let w be a vertex of $K^{(1)}$. Then

$$\psi_0(1 \cdot \langle w \rangle) = 1 \cdot \langle v \rangle$$

where v is a vertex of some simplex σ of K of which w is the barycenter. Then

$$\varphi_0\psi_0(1 \cdot \langle w \rangle) = \varphi_0(1 \cdot \langle v \rangle) = 1 \cdot \langle v \rangle.$$

Thus

$$1 \cdot \langle w \rangle - \varphi_0\psi_0(1 \cdot \langle w \rangle) = 1 \cdot \langle w \rangle - 1 \cdot \langle v \rangle = \partial(1 \cdot \langle vw \rangle),$$

so we define

$$D_0(1 \cdot \langle w \rangle) = 1 \cdot \langle vw \rangle.$$

The function D_0 is defined by this procedure for every elementary 0-chain $1 \cdot \langle w \rangle$ and extended by linearity to a homomorphism $D_0: C_0(K^{(1)}) \to C_1(K^{(1)})$. Proceeding inductively, suppose that $D_0, \ldots, D_{p-1}$ have all been defined, and let $1 \cdot \tau^p$ be an elementary p-chain on $K^{(1)}$. Then, for every $(p-1)$-chain c,

$$c - \varphi_{p-1}\psi_{p-1}(c) = \partial D_{p-1}(c) + D_{p-2}\partial(c),$$

so

$$\partial D_{p-1}(c) = c - \varphi_{p-1}\psi_{p-1}(c) - D_{p-2}\partial c.$$

Consider

$$z = 1 \cdot \tau^p - \varphi_p\psi_p(1 \cdot \tau^p) - D_{p-1}\partial(1 \cdot \tau^p).$$

Then

$$\begin{aligned}
\partial z &= \partial(1 \cdot \tau^p) - \partial\varphi_p\psi_p(1 \cdot \tau^p) - \partial D_{p-1}\partial(1 \cdot \tau^p) \\
&= \partial(1 \cdot \tau^p) - \varphi_{p-1}\psi_{p-1}\partial(1 \cdot \tau^p) \\
&\quad - (\partial(1 \cdot \tau^p) - \varphi_{p-1}\psi_{p-1}\partial(1 \cdot \tau^p) - D_{p-2}\partial\partial(1 \cdot \tau^p)) = 0.
\end{aligned}$$

This means that z is a cycle on $K^{(1)}$. An argument analogous to that used in the proof of Theorem 2.9 shows that z is the boundary of a $(p + 1)$-chain c_{p+1} on $K^{(1)}$. We then define

$$D_p(1 \cdot \tau^p) = c_{p+1}$$

and extend by linearity. This completes the definition of the deformation operator $\mathscr{D}$ and shows that K and $K^{(1)}$ are chain equivalent. □

Theorem 7.6. *The homology groups $H_p(K)$ and $H_p(K^{(n)})$ are isomorphic for all integers $p \geq 0$, $n \geq 1$, and each complex K.*

Proof. The inductive definition of $K^{(n)}$ and Theorem 7.5 show that K and $K^{(n)}$ are chain equivalent for $n \geq 1$. Theorem 7.4 then shows that $H_p(K) \cong H_p(K^{(n)})$, $p \geq 0$. □

Deformation operators were invented by Solomon Lefschetz (1884–1972). The proof of Theorem 7.5 given above is due to Lefschetz [13, 15].

Let $|K|$ and $|L|$ be polyhedra with triangulations K and L respectively and $f: |K| \to |L|$ a continuous map. We now have the machinery necessary to prove that the induced homomorphisms $f_p^*: H_p(K) \to H_p(L)$ are uniquely determined by f. Recall that this problem was postponed in Chapter 3. According to the Simplicial Approximation Theorem (Theorem 3.6), there is a barycentric subdivision $K^{(k)}$ of K and a simplicial mapping g from $K^{(k)}$ to L such that, as functions from $|K|$ to $|L|$, f and g are homotopic. There is some freedom in the choices of g and the degree k of the barycentric subdivision. From the proof of Theorem 3.6, k must be large enough so that $K^{(k)}$ is star related to L relative to f. The simplicial map g is given by the proof of Theorem 3.4; for a vertex u of $K^{(k)}$, $g(u)$ may be any vertex of L satisfying

$$f(\text{ost}(u)) \subset \text{ost}(g(u)).$$

To show that the sequence of homomorphisms is independent of the admissible choices for g, it is sufficient to prove that any admissible change in the value of g at one vertex does not alter the induced homomorphisms $g_p^*: H_p(K^{(k)}) \to H_p(L)$. Any simplicial map satisfying the requirements of Theorem 3.4 can be obtained from any other one by a finite sequence of such changes at single vertices. Suppose then that g and h are two simplicial mappings from $K^{(k)}$ into L which have identical values at each vertex of $K^{(k)}$ except for one vertex v and that, for this vertex, $\text{ost}(g(v))$ and $\text{ost}(h(v))$ both contain $f(\text{ost}(v))$. We shall show that the chain mappings $\{g_p: C_p(K^{(k)}) \to C_p(L)\}$ and $\{h_p: C_p(K^{(k)}) \to C_p(L)\}$ are chain homotopic and conclude from Theorem 7.3 that the induced homomorphisms g_p^* and h_p^* from $H_p(K^{(k)})$ to $H_p(L)$ are identical for each value of p.

For our deformation operator $\mathscr{D} = \{D_p: C_p(K^{(k)}) \to C_{p+1}(L)\}_{-1}^{\infty}$, we must have $D_{-1} = 0$. For any vertex u of $K^{(k)}$ with $u \neq v$, define $D_0(1 \cdot \langle u \rangle) = 0$, and define

$$D_0(1 \cdot \langle v \rangle) = 1 \cdot \langle h(v) g(v) \rangle.$$

Now extend D_0 by linearity to a homomorphism from $C_0(K^{(k)})$ to $C_1(L)$.

Note that

$$\partial D_0(1\cdot\langle v\rangle) + D_{-1}\partial(1\cdot\langle v\rangle) = \partial(1\cdot\langle h(v)g(v)\rangle) = 1\cdot\langle g(v)\rangle - 1\cdot\langle h(v)\rangle$$
$$= g_0(1\cdot\langle v\rangle) - h_0(1\cdot\langle v\rangle).$$

If u is a vertex of $K^{(k)}$ different from v, then

$$g_0(1\cdot\langle u\rangle) = h_0(1\cdot\langle u\rangle), \qquad D_0(1\cdot\langle u\rangle) = 0,$$

so the desired relation

$$\partial D_p + D_{p-1}\partial = g_p - h_p$$

holds for $p = 0$.

For the general case, let $1\cdot\sigma^p$ be an elementary p-chain in $C_p(K^{(k)})$. If v is not a vertex of σ^p, then we define $D_p(1\cdot\sigma^p) = 0$ in $C_{p+1}(L)$. If v is a vertex of σ^p, then $\sigma^p = v\sigma^{p-1}$ for some $(p-1)$-simplex σ^{p-1}, and we define

$$D_p(1\cdot\sigma^p) = 1\cdot h(v)g(v)\tau$$

where τ is the $(p-1)$-simplex in L which is the image of σ^{p-1} under both g and h. As usual, D_p is extended linearly to a homomorphism from $C_p(K^{(k)})$ to $C_{p+1}(L)$. Then for the case in which v is a vertex of σ^p,

$$\begin{aligned}
\partial D_p(1\cdot\sigma^p) &+ D_{p-1}\partial(1\cdot\sigma^p)\\
&= \partial(1\cdot h(v)g(v)\tau) + D_{p-1}\partial(1\cdot v\sigma^{p-1})\\
&= 1\cdot g(v)\tau - h(v)\partial(1\cdot g(v)\tau) + D_{p-1}(1\cdot\sigma^{p-1} - v\partial(1\cdot\sigma^{p-1}))\\
&= 1\cdot g(v)\tau - h(v)[1\cdot\tau - g(v)\partial(1\cdot\tau)] - D_{p-1}(v\partial(1\cdot\sigma^{p-1}))\\
&= 1\cdot g(v)\tau - 1\cdot h(v)\tau + h(v)g(v)\partial(1\cdot\tau) - h(v)g(v)\partial(1\cdot\tau)\\
&= g_p(1\cdot v\sigma^{p-1}) - h_p(1\cdot v\sigma^{p-1}) = g_p(1\cdot\sigma^p) - h_p(1\cdot\sigma^p).
\end{aligned}$$

Thus

$$\partial D_p + D_{p-1}\partial = g_p - h_p, \qquad p \geq 0,$$

and the chain mappings induced by g and h must be chain homotopic. Theorem 7.3 now guarantees that $g_p^* = h_p^*$, so we conclude that the induced homomorphism f_p^* is independent of the allowable choices of the simplicial map g.

Question: Where did we use the assumption that $\text{ost}(g(v))$ and $\text{ost}(h(v))$ both contain $f(\text{ost}(v))$?

The homomorphism $f_p^*: H_p(K) \to H_p(L)$ is actually the composition $g_p^*\mu_p^*$ from the diagram

$$H_p(K) \xrightarrow{\mu_p^*} H_p(K^{(k)}) \xrightarrow{g_p^*} H_p(L)$$

where μ_p^* is the isomorphism induced by chain derivation. For a barycentric subdivision $K^{(r)}$ of higher degree, let $\psi_p^*: H_p(K) \to H_p(K^{(r)})$ be the isomorphism induced by chain derivation and $j_p^*: H_p(K^{(r)}) \to H_p(L)$ the homomorphism induced by an admissible simplicial map. It is left as an exercise for the reader to show that $g_p^*\mu_p^* = j_p^*\psi_p^*$ and hence that f_p^* is also independent of the allowable choices for the degree of the barycentric subdivision $K^{(k)}$.

7.2 The Lefschetz Fixed Point Theorem

This section is devoted to the most famous of all the theorems about fixed points of continuous maps. Lefschetz introduced in 1926 a number $\lambda(f)$ associated with each continuous map $f: |K| \to |K|$ from a polyhedron into itself. If the Lefschetz number $\lambda(f)$ is not zero, then f has at least one fixed point. (The Lefschetz number does not specify the number of fixed points.) Brouwer's Fixed Point Theorem (Theorem 3.13) can be proved as a simple corollary.

In this section we assume that rational numbers rather than integers are used as the coefficient group for chains. Thus the pth chain group $C_p(K)$ of a complex K is considered a vector space over the field of rational numbers.

Definition. Let K be a complex with $\{\sigma_i^p\}$ its set of p-simplexes, and let $\varphi = \{\varphi_p\}$ be a chain mapping on K. For a p-simplex σ_i^p of K,

$$\varphi_p(1 \cdot \sigma_i^p) = \sum_{\sigma_j^p \in K} a_{ij}^p \cdot \sigma_j^p$$

for some rational numbers a_{ij}^p, one for each p-simplex σ_j^p of K. Then σ_i^p is a *fixed simplex* of φ provided that a_{ii}^p, the coefficient of σ_i^p in the expansion of $\varphi_p(1 \cdot \sigma_i^p)$, is not zero. The number $(-1)^p a_{ii}^p$ is called the *weight* of the fixed simplex σ_i^p. Let

$$A_p = (a_{ij}^p)$$

be the matrix whose entry in row i and column j is a_{ij}^p. Since the trace of a square matrix is the sum of its diagonal elements, then

$$\text{trace } A_p = \sum a_{ii}^p,$$

and the number

$$\lambda(\varphi) = \sum_p (-1)^p \text{ trace}(A_p)$$

is the sum of the weights of all the fixed simplexes of φ. The number $\lambda(\varphi)$ is called the *Lefschetz number* of φ. (Note that if $\lambda(\varphi) \neq 0$, then φ must have at least one fixed simplex in some dimension p.)

The matrix $A_p = (a_{ij}^p)$ is the matrix of φ_p as a linear transformation from the vector space $C_p(K)$ into itself relative to the basis of elementary p-chains $\{1 \cdot \sigma_i^p\}$. Since the trace of the matrix of a linear transformation is not affected by a change of basis, the Lefschetz number $\lambda(\varphi)$ is the same regardless of the choice of basis for $C_p(K)$.

Example 7.3. Let $\varphi_p: C_p(K) \to C_p(K)$ be the identity map on $C_p(K)$ for some complex K, $p \geq 0$. Then

$$a_{ii}^p = 1, \qquad a_{ij}^p = 0 \quad \text{for } i \neq j,$$

and each simplex is a fixed simplex. Thus

$$\lambda(\varphi) = \sum (-1)^p \text{ trace } A_p = \sum (-1)^p \alpha_p = \chi(K)$$

where α_p is the number of simplexes of dimension p and $\chi(K)$ is the Euler characteristic of K. Thus the Lefschetz number is a generalization of the Euler characteristic.

Theorem 7.7. *Let $\varphi = \{\varphi_p\}$ be a chain mapping on a complex K. The Lefschetz number $\lambda(\varphi)$ is completely determined by the induced homomorphisms $\varphi_p^*: H_p(K) \to H_p(K)$ on the homology groups.*

Proof. The proof is similar to the proof of the Euler–Poincaré Theorem (Theorem 2.5), and we use the same notation. Then $\{z_p^i\} \cup \{b_p^i\}$ is a basis for the cycle vector space Z_p, $\{b_p^i\}$ is a basis for the boundary space B_p, $\{d_p^i\}$ is a basis for D_p, $b_p^i = \partial d_{p+1}^i$, and n is the dimension of K, as in the proof of Theorem 2.5. Note that $\{b_p^i\} \cup \{z_p^i\} \cup \{d_p^i\}$ is a basis for C_p. For any b_p^i,

$$\varphi_p(b_p^i) = \sum_j a_{ij}^p b_p^j, \qquad 0 \leq p \leq n-1,$$

for some rational coefficients a_{ij}^p since the linear transformation φ_p takes B_p into B_p. For any z_p^i, $0 \leq p \leq n$, $\varphi_p(z_p^i)$ must be a cycle, so there are coefficients $a_{ij}'^p$, e_{ij}^p such that

$$\varphi_p(z_p^i) = \sum_j a_{ij}'^p b_p^j + \sum_j e_{ij}^p z_p^j.$$

For any d_p^i, $1 \leq p \leq n$, there are coefficients $a_{ij}''^p$, $e_{ij}'^p$, g_{ij}^p such that

$$\varphi_p(d_p^i) = \sum_j a_{ij}''^p b_p^j + \sum_j e_{ij}'^p z_p^j + \sum_j g_{ij}^p d_p^j.$$

Then

$$\lambda(\varphi) = \sum_{i=0}^{n} (-1)^p(\text{trace } A_p + \text{trace } E_p + \text{trace } G_p)$$

where

$$A_p = (a_{ij}^p), \qquad E_p = (e_{ij}^p), \qquad G_p = (g_{ij}^p),$$

and $A_n = G_0$ is the zero matrix. Now

$$\partial\varphi_{p+1}(d_{p+1}^i) = \varphi_p\partial(d_{p+1}^i) = \varphi_p(b_p^i) = \sum a_{ij}^p b_p^j.$$

Also,

$$\partial\varphi_{p+1}(d_{p+1}^i) = \partial\left(\sum a_{ij}''^{p+1} b_{p+1}^j + \sum e_{ij}'^{p+1} z_{p+1}^j + \sum g_{ij}^{p+1} d_{p+1}^j\right)$$
$$= \sum g_{ij}^{p+1}\partial(d_{p+1}^j) = \sum g_{ij}^{p+1} b_p^j.$$

Then

$$a_{ij}^p = g_{ij}^{p+1}, \qquad A_p = G_{p+1}, \qquad 0 \leq p \leq n-1,$$

and the sum

$$\lambda(\varphi) = \sum_{i=0}^{n} (-1)^p(\text{trace } A_p + \text{trace } E_p + \text{trace } G_p)$$

telescopes to give

$$\lambda(\varphi) = \sum_{i=0}^{n} (-1)^p \text{ trace } E_p.$$

This means that the Lefschetz number $\lambda(\varphi)$ is completely determined by the action of the maps φ_p on the generating cycles z_p^i of $H_p(K)$. The coefficients e_{ij}^p are determined by the induced homomorphisms $\varphi_p^*: H_p(K) \to H_p(K)$ because the homology classes $[z_p^i]$ generate $H_p(K)$:

$$\varphi_p^*([z_p^i]) = \sum_j e_{ij}^p[z_p^j].$$

Thus the induced homomorphisms completely determine the coefficients e_{ij}^p which completely determine $\lambda(\varphi)$, so the theorem follows. □

Thus far we have defined the Lefschetz number for chain mappings. This definition must be extended to continuous mappings.

Definition. Let K be a complex and $f: |K| \to |K|$ a continuous function. Let $K^{(s)}$ be a barycentric subdivision of K and g a simplicial map from $K^{(s)}$ to K which is a simplicial approximation of f (Theorem 3.6). Then g induces a chain mapping $\{g_p: C_p(K^{(s)}) \to C_p(K)\}$. Let $\mu = \{\mu_p: C_p(K) \to C_p(K^{(s)})\}$ be the sth chain derivation on K. The *Lefschetz number* $\lambda(f)$ of f is the Lefschetz number of the composite chain mapping $\{g_p\mu_p: C_p(K) \to C_p(K)\}$.

It appears that the Lefschetz number is influenced by the possible choices for g and s. The number is independent of these choices, however, since it is completely determined by the induced homomorphisms

$$f_p^* = g_p^*\mu_p^*: H_p(K) \to H_p(K)$$

and f_p^* is independent of the allowable choices for g and s.

Theorem 7.8 (The Lefschetz Fixed Point Theorem). *Let K be a complex and $f: |K| \to |K|$ a continuous map. If the Lefschetz number $\lambda(f)$ is not* 0, *then f has a fixed point.*

PROOF. Suppose to the contrary that f has no fixed point. Since $|K|$ is compact, there is a number $\epsilon > 0$ such that if $x \in |K|$, then the distance $\|f(x) - x\| \geq \epsilon$. By replacing K with a suitable barycentric subdivision if necessary, we may assume that mesh $K < \epsilon/3$. According to the proof of the Simplicial Approximation Theorem (Theorem 3.6), there is a positive integer s and a simplicial map g from $K^{(s)}$ to K homotopic to f such that, for each x in $|K|$, $f(x)$ and $g(x)$ lie in a common simplex of K. Then $\|f(x) - g(x)\| < \epsilon/3$ for all $x \in |K|$.

Suppose that some simplex σ of K contains a point x such that $g(x)$ is also in σ. Then

$$\|f(x) - x\| \leq \|f(x) - g(x)\| + \|g(x) - x\| < 2\epsilon/3,$$

which contradicts the fact that $\|f(x) - x\| \geq \epsilon$. Thus σ and $g(\sigma)$ are disjoint for all σ in K. Consider the sth chain derivation $\mu = \{\mu_p: C_p(K) \to C_p(K^{(s)})\}$ and the chain mapping $\{g_p: C_p(K^{(s)} \to C_p(K)\}$ induced by g. If σ^p is a p-simplex of K, then $\mu_p(1 \cdot \sigma^p)$ is a chain on $K^{(s)}$ all of whose simplexes with nonzero coefficient are contained in σ^p. Since σ^p and $g(\sigma^p)$ are disjoint, then $g_p\mu_p(1 \cdot \sigma^p)$ is a p-chain on K none of whose simplexes with nonzero coefficient intersects σ. Thus $g_p\mu_p$ has no fixed simplex, and the Lefschetz number of the chain mapping $\{g_p\mu_p\}$ is zero. But this is the Lefschetz number of f, contradicting the hypothesis $\lambda(f) \neq 0$. □

Corollary (The Brouwer Fixed Point Theorem). *If σ^n is an n-simplex, n a positive integer, and $f\colon \sigma^n \to \sigma^n$ a continuous map, then f has a fixed point.*

PROOF. Let $K = \mathrm{Cl}(\sigma^n)$. Then $H_0(K) \cong \mathbb{Z}$, $H_p(K) = \{0\}$ for $p > 0$. Let v be a vertex of σ^n so that the homology class $[1 \cdot \langle v \rangle]$ may be considered a generator of $H_0(K)$ (Theorem 2.4). Then

$$f_0^*([1 \cdot \langle v \rangle]) = [1 \cdot \langle v \rangle],$$

and the coefficient matrix E_0 of Theorem 7.7 has trace 1. (Why?) Each matrix E_p for $p > 0$ has only zero entries, and hence

$$\lambda(f) = \sum (-1)^p \operatorname{trace} E_p = 1.$$

Thus $\lambda(f) \neq 0$, so f must have a fixed point. □

Corollary. *Every continuous map from S^n to S^n, $n \geq 1$, whose degree is not 1 or -1 has a fixed point.*

PROOF. Recall from Theorem 2.9 that $H_0(S^n) \cong H_n(S^n) \cong \mathbb{Z}$ and $H_p(S^n) = \{0\}$ otherwise. If $[1 \cdot \langle v \rangle]$ and $[z_n]$ are generators of $H_0(S^n)$ and $H_n(S^n)$ respectively, then

$$f_0^*([1 \cdot \langle v \rangle]) = [1 \cdot \langle v \rangle],$$
$$f_n^*([z_n]) = d[z_n]$$

where d is the degree of f. Then

$$\lambda(f) = 1 + (-1)^n d,$$

so $\lambda(f) \neq 0$ if d is not 1 or -1. □

Corollary. *If $f\colon S^n \to S^n$ is the antipodal map, then the degree of f is $(-1)^{n+1}$.*

PROOF. Since f has no fixed point, then $\lambda(f) = 0$. Hence

$$0 = 1 + (-1)^n d$$

where d is the degree of f. This gives $d = (-1)^{n+1}$. □

The Lefschetz Fixed Point Theorem was discovered by Lefschetz in 1926 [47, 48]. A simpler proof, the one used in this book, was published by H. Hopf in 1928 [40].

7.3 Relative Homology Groups

Suppose that K is a complex and L is a complex contained in K. It often happens that one knows the homology groups of either K of L and needs to know the homology groups of the other. The groups $H_p(K)$ and $H_p(L)$ can be compared using the "relative homology groups" $H_p(K/L)$ to which this section is devoted. The intuitive idea is to "remove" all chains on L by considering quotient groups. The groups $H_p(K)$, $H_p(L)$, and $H_p(K/L)$ form a sequence of groups and homomorphisms called the "homology sequence." Using this sequence, one can often compute any one of the groups $H_p(K)$, $H_p(L)$, or $H_p(K/L)$ provided that enough information is known about the others.

Definition. A *subcomplex* of a complex K is a complex L with the property that each simplex of L is a simplex of K.

Note that not every subset of a complex is a subcomplex; the subset must be a complex in its own right. The p-skeleton of a complex is one type of subcomplex. Note also that the empty set $\varnothing$ is a subcomplex of each complex K; the relative homology groups $H_p(K/L)$ will reduce to $H_p(K)$ when $L = \varnothing$.

Definition. Let K be a complex with subcomplex L. By assigning value 0 to each simplex of the complement $K\backslash L$, each chain on L can be considered a chain on K, and we can consider $C_p(L)$ as a subgroup of $C_p(K)$, $p \geq 0$. The *relative p-dimensional chain group of K modulo L*, or *relative p-chain group* (with integer coefficients), is the quotient group

$$C_p(K/L) = C_p(K)/C_p(L).$$

Thus each member of $C_p(K/L)$ is a coset $c_p + C_p(L)$ where $c_p \in C_p(K)$.

For $p \geq 1$, the *relative boundary operator*

$$\partial: C_p(K/L) \to C_{p-1}(K/L)$$

is defined by

$$\partial(c_p + C_p(L)) = \partial c_p + C_{p-1}(L), \qquad (c_p + C_p(L)) \in C_p(K/L),$$

where ∂c_p denotes the usual boundary of the p-chain c_p. It is easily observed that the relative boundary operator is a homomorphism.

The *group of relative p-dimensional cycles on K modulo L*, denoted by $Z_p(K/L)$, is the kernel of the relative boundary operator

$$\partial: C_p(K/L) \to C_{p-1}(K/L), \qquad p \geq 1.$$

We define $Z_0(K/L)$ to be the chain group $C_0(K/L)$.

For $p \geq 0$, the *group of relative p-dimensional boundaries on K modulo L*, denoted by $B_p(K/L)$, is the image $\partial(C_{p+1}(K/L))$ of $C_{p+1}(K/L)$ under the relative boundary homomorphism.

The *relative p-dimensional simplicial homology group of K modulo L* is the quotient group

$$H_p(K/L) = \frac{Z_p(K/L)}{B_p(K/L)}, \qquad p \geq 0.$$

In order for the homology group $H_p(K/L)$ to make sense, every relative p-boundary must be a relative p-cycle. In other words, we must have $B_p(K/L)$

$\subset Z_p(K/L)$ for the quotient group to be defined. The verification of this fact is left as an easy exercise

The members of $H_p(K/L)$ are denoted $[z_p + C_p(L)]$ where $z_p + C_p(L)$ is a relative p-cycle. It is required that ∂z_p be a $(p - 1)$-chain on L, not that z_p be an actual cycle. However, if z_p is a cycle, then $z_p + C_p(L)$ is certainly a relative cycle.

Example 7.4. Let K be the 1-skeleton of a 2-simplex $\langle v_0v_1v_2\rangle$ and L the subcomplex determined by the vertex v_0. Let us determine $H_0(K/L)$ and $H_1(K/L)$. For the case $p = 0$,

$$C_0(K) = Z_0(K) \cong \mathbb{Z} \oplus \mathbb{Z} \oplus \mathbb{Z},$$
$$C_0(L) = Z_0(L) \cong \mathbb{Z}, \qquad C_0(K/L) = Z_0(K/L) \cong \mathbb{Z} \oplus \mathbb{Z}.$$

The members of $Z_0(K/L)$ are chains of the form

$$z = g_1\cdot\langle v_1\rangle + g_2\cdot\langle v_2\rangle + C_0(L), \qquad g_1, g_2 \in \mathbb{Z},$$

where

$$C_0(L) = \{g\cdot\langle v_0\rangle : g \text{ is an integer}\}.$$

But

$$\partial(g_1\cdot\langle v_0v_1\rangle + g_2\cdot\langle v_0v_2\rangle) = g_1\cdot\langle v_1\rangle + g_2\cdot\langle v_2\rangle + (-g_1 - g_2)\cdot\langle v_0\rangle,$$

so

$$\partial(g_1\cdot\langle v_0v_1\rangle + g_2\cdot\langle v_0v_2\rangle + C_1(L)) = g_1\cdot\langle v_1\rangle + g_2\cdot\langle v_2\rangle + C_0(L).$$

Thus every relative 0-cycle is a relative 0-boundary. This means that

$$Z_0(K/L) = B_0(K/L), \qquad H_0(K/L) = \{0\}.$$

Now suppose $p = 1$. Let

$$w = h_1\cdot\langle v_0v_1\rangle + h_2\cdot\langle v_1v_2\rangle + h_3\cdot\langle v_0v_2\rangle + C_1(L)$$

be a relative 1-chain. (Since $C_1(L) = \{0\}$, 1-chains and relative 1-chains are essentially the same.) Then

$$\partial w = (h_1 - h_2)\cdot\langle v_1\rangle + (h_2 + h_3)\cdot\langle v_2\rangle + C_0(L).$$

Then w is a relative 1-cycle if and only if $h_1 = h_2 = -h_3$. Hence $Z_1(K/L) \cong \mathbb{Z}$. Since K has no 2-simplexes, then $B_1(K/L) = \{0\}$ and $H_1(K/L) \cong \mathbb{Z}$. Since there are no simplexes of dimension 2 or higher, then $H_p(K/L) = \{0\}$, $p \geq 2$.

Example 7.5. Let K denote the closure of a 2-simplex $\sigma^2 = \langle v_0v_1v_2\rangle$ and L its 1-skeleton. Since K and L have precisely the same 0-simplexes and 1-simplexes, then

$$C_0(K) = C_0(L), \qquad C_0(K/L) = \{0\}, \qquad H_0(K/L) = \{0\},$$
$$C_1(K) = C_1(L), \qquad C_1(K/L) = \{0\}, \qquad H_1(K/L) = \{0\}.$$

Since L has no simplexes of dimension two or higher, it might appear at first that $H_p(K)$ and $H_p(K/L)$ are isomorphic for $p \geq 2$. This is true for $p \geq 3$ but not for $p = 2$. Although L has no simplexes of dimension two, it does affect $Z_2(K/L)$. The reason is that the boundary of a 2-chain is a 1-chain; if the 1-chain has nonzero coefficients only for simplexes of L, then the 2-chain is a relative cycle. In this case, the elementary relative 2-chain

$$u = g\cdot\langle v_0v_1v_2\rangle + C_2(L), \qquad g \in \mathbb{Z},$$

has relative boundary

$$\partial u = g\cdot\langle v_1v_2\rangle - g\cdot\langle v_0v_2\rangle + g\cdot\langle v_0v_1\rangle + C_1(L) = 0$$

because all 1-simplexes of K are in L. Thus the subcomplex L produces relative 2-cycles, and $Z_2(K/L) \cong \mathbb{Z}$. Since $B_2(K/L) = \{0\}$, then $H_2(K/L) \cong \mathbb{Z}$. Note in particular that $H_2(K) = \{0\}$, so $H_2(K/L)$ is not isomorphic to $H_2(K)$.

Our next objective is to show that there is a special sequence

$$\cdots \xrightarrow{\partial^*} H_p(L) \xrightarrow{i^*} H_p(K) \xrightarrow{j^*} H_p(K/L) \xrightarrow{\partial^*} H_{p-1}(L) \xrightarrow{i^*} \cdots \xrightarrow{i^*} H_0(K) \xrightarrow{j^*} H_0(K/L)$$

where i^*, j^*, and ∂^* are homomorphisms. Strictly speaking, each homomorphism should be marked by p, indicating the dimension, but this notation is cumbersome. The dimension will always be known from the subscripts on the homology groups.

Definition. Let K be a complex with subcomplex L. The inclusion map i from L into K is simplicial and induces a homomorphism $i^*\colon H_p(L) \to H_p(K)$ for each $p \geq 0$. The effect of this homomorphism is easily described: If $[z_p] \in H_p(L)$ is represented by the p-cycle z_p on L, then z_p can be considered a p-cycle on K. Then z_p determines a homology class $i^*([z_p]) = [z_p]$ in $H_p(K)$.

Let $j\colon C_p(K) \to C_p(K/L)$ be the homomorphism defined by

$$j(c_p) = c_p + C_p(L), \qquad c_p \in C_p(K).$$

Then j induces a homomorphism $j^*\colon H_p(K) \to H_p(K/L)$, $p \geq 0$. If $[z_p] \in H_p(K)$, then $z_p + C_p(L)$ is a relative p-cycle and determines a member $[z_p + C_p(L)]$ of $H_p(K/L)$. The homomorphism j^* takes $[z_p]$ to $[z_p + C_p(L)]$.

The definition of $\partial^*\colon H_p(K/L) \to H_{p-1}(L)$ comes next. If $[z_p + C_p(L)] \in H_p(K/L)$, $p \geq 1$, then $z_p + C_p(L)$ is a relative p-cycle. This means that ∂z_p is in $C_{p-1}(L)$. Since $\partial\partial z_p = 0$, then ∂z_p is a $(p-1)$-cycle on L and determines a member $[\partial z_p]$ of $H_{p-1}(L)$. We define

$$\partial^*([z_p + C_p(L)]) = [\partial z_p], \qquad [z_p + C_p(L)] \in H_p(K/L).$$

The *homology sequence* of the pair (K, L) is the sequence of groups and homomorphisms

$$\cdots \xrightarrow{\partial^*} H_p(L) \xrightarrow{i^*} H_p(K) \xrightarrow{j^*} H_p(K/L) \xrightarrow{\partial^*} H_{p-1}(L) \xrightarrow{i^*} \cdots \xrightarrow{i^*} H_0(K) \xrightarrow{j^*} H_0(K/L).$$

The reader is asked to verify that i^*, j^*, and ∂^* are well-defined homomorphisms. The homology sequence has a nice algebraic structure whose basic properties are developed in the next definition and the two theorems that follow it.

Definition. A sequence

$$\cdots \xrightarrow{h_{p+1}} G_p \xrightarrow{h_p} G_{p-1} \xrightarrow{h_{p-1}} \cdots \xrightarrow{h_2} G_1 \xrightarrow{h_1} G_0$$

of groups $G_0, G_1, \ldots$ and homomorphisms $h_1, h_2, \ldots$ is *exact* provided that the kernel of h_{p-1} equals the image $h_p(G_p)$ for $p \geq 2$ and that h_1 maps G_1 onto G_0. (Requiring that h_1 be onto is equivalent to requiring that G_0 be followed by the trivial group.)

There are many theorems that compare the groups of an exact sequence. The following is the simplest.

Theorem 7.9. *Suppose that an exact sequence has a section of four groups*

$$\{0\} \xrightarrow{f} A \xrightarrow{g} B \xrightarrow{h} \{0\}$$

where $\{0\}$ *denotes the trivial group. Then* g *is an isomorphism from* A *onto* B.

PROOF. The image $f(\{0\}) = \{0\}$ contains only the identity element of A. Exactness then guarantees that g has kernel $\{0\}$, so g is one-to-one. The kernel of h is all of B, and this must be the image $g(A)$. Thus g is an isomorphism as claimed. □

Theorem 7.10. *Suppose that an exact sequence has a section of five groups*

$$\{0\} \to A \xrightarrow{f} B \xrightarrow{g} C \to \{0\},$$

there is a homomorphism $h: C \to B$ *such that* gh *is the identity map on* C, *and* B *is abelian. Then* $B \simeq A \oplus C$.

It is left as an exercise for the reader to show that $T: A \oplus C \to B$ defined by

$$T(a, c) = f(a) \cdot h(c), \qquad (a, c) \in A \oplus C,$$

is the required isomorphism.

Theorem 7.11. *If* K *is a complex with subcomplex* L, *then the homology sequence of* (K, L) *is exact.*

PROOF. In the homology sequence

$$\cdots \xrightarrow{\partial^*} H_p(L) \xrightarrow{i^*} H_p(K) \xrightarrow{j^*} H_p(K/L) \xrightarrow{\partial^*} H_{p-1}(L) \xrightarrow{i^*} \cdots \xrightarrow{i^*} H_0(K) \xrightarrow{j^*} H_0(K/L),$$

we must show that the last homomorphism j^* maps $H_0(K)$ onto $H_0(K/L)$ and that the kernel of each homomorphism is the image of the one that precedes it.

To see that j^* is onto, let $[z_0 + C_0(L)] \in H_0(K/L)$. Then z_0 is a 0-chain on K, and

$$j^*[z_0] = [z_0 + C_0(L)],$$

so j^* is onto.

The remainder of the proof breaks naturally into six parts:

(1) image $i^* \subset$ kernel j^*,
(2) kernel $j^* \subset$ image i^*,
(3) image $j^* \subset$ kernel ∂^*,
(4) kernel $\partial^* \subset$ image j^*,
(5) image $\partial^* \subset$ kernel i^*,
(6) kerenel $i^* \subset$ image ∂^*.

To prove (1), let $i^*([z_p])$ be in the image of i^* where z_p is a p-cycle on L. Then

$$j^*i^*([z_p]) = [z_p + C_p(L)] = [0 + C_p(L)] = 0$$

since $z_p \in C_p(L)$. Thus image $i^* \subset$ kernel j^*.

For part (2), let $[w_p] \in H_p(K)$ be an element of the kernel of j^*; $j^*([w_p]) = 0$ in $H_p(K/L)$. We must find an element $[z_p]$ in $H_p(L)$ such that $i^*([z_p]) = [w_p]$. Since

$$j^*([w_p]) = [w_p + C_p(L)] = 0,$$

then $w_p + C_p(L)$ is the relative boundary of a relative $(p + 1)$-chain $c_{p+1} + C_{p+1}(L)$:

$$\partial c_{p+1} + C_p(L) = w_p + C_p(L),$$

so $w_p - \partial c_{p+1}$ is in $C_p(L)$. Since both w_p and ∂c_{p+1} are cycles on K, then $w_p - \partial c_{p+1}$ is also a cycle and determines a member $[w_p - \partial c_{p+1}]$ of $H_p(L)$. Note that

$$i^*([w_p - \partial c_{p+1}]) = [w_p - \partial c_{p+1}] = [w_p]$$

since w_p and $w_p - \partial c_{p+1}$ are homolgous cycles on K. Thus kernel $j^* \subset$ image i^*.

For part (3), let $j^*([z_p]) = [z_p + C_p(L)]$ be a member of the image of j^* where z_p is a p-cycle on K. Then

$$\partial^*j^*([z_p]) = \partial^*([z_p + C_p(L)]) = [\partial z_p] = 0$$

since $\partial z_p = 0$. Thus image $j^* \subset$ kernel ∂^*.

Proceeding to (4), let $[x_p + C_p(L)]$ be in the kernel of ∂^* where $x_p + C_p(L)$ is a relative p-cycle. Then

$$\partial^*([x_p + C_p(L)]) = [\partial x_p] = 0$$

in $H_{p-1}(L)$. This means that

$$\partial x_p = \partial y_p$$

for some p-chain y_p on L. Then $x_p - y_p$ is a p-cycle on K and determines a member $[x_p - y_p]$ of $H_p(K)$. Note that

$$j^*([x_p - y_p]) = [x_p - y_p + C_p(L)] = [x_p + C_p(L)]$$

since $y_p \in C_p(L)$. Thus $[x_p + C_p(L)]$ is in the image of j^*, so kernel $\partial^* \subset$ image j^*.

Parts (5) and (6) are left to the reader. □

Example 7.6. Let K denote the closure of an n-simplex and L its $(n-1)$-skeleton, $n \geq 2$. We shall use the homology sequence to compute $H_p(K/L)$ thus generalizing Example 7.5.

Since $n \geq 2$, K and L have the same 0-chains and the same 1-chains, and

$$H_0(K/L) = H_1(K/L) = \{0\}.$$

For $p > 1$, consider the homology sequence

$$\cdots \to H_p(K) \to H_p(K/L) \to H_{p-1}(L) \to H_{p-1}(K) \to \cdots$$

Since $H_{p-1}(K) = H_p(K) = \{0\}$, Theorem 7.9 shows that $H_p(K/L) \cong H_{p-1}(L)$, $p > 1$. Since $|L|$ is homeomorphic to S^{n-1}, then

$$H_n(K/L) \cong H_{n-1}(S^{n-1}) \cong \mathbb{Z},$$

and $H_p(K/L) = \{0\}$ if $p \neq n$.

Example 7.7. Let X be the union of two n-spheres tangent at a point. Then X has as triangulation the n-skeleton of the closure of two $(n+1)$-simplexes joined at a common vertex. Denote this triangulation by K, and let L denote the n-skeleton of one of the two $(n+1)$-simplexes. The section

$$H_{n+1}(K/L) \xrightarrow{\partial^*} H_n(L) \xrightarrow{i^*} H_n(K) \xrightarrow{j^*} H_n(K/L) \xrightarrow{\partial^*} H_{n-1}(L)$$

of the homology sequence of (K, L) satisfies the hypotheses of Theorem 7.10 so that

$$H_n(K) \cong H_n(K/L) \oplus H_n(L).$$

The reader should show that

$$H_n(K/L) \cong H_n(L) \cong \mathbb{Z}$$

and

$$H_n(X) = H_n(K) \cong \mathbb{Z} \oplus \mathbb{Z}.$$

The relative homology groups were defined by Lefschetz [46] in 1927, and the homology sequence was introduced by Hurewicz [43] in 1941. The six parts of the exactness argument (Theorem 7.11) had been used separately for many years before Hurewicz' formalization of the homology sequence, however.

7.4 Singular Homology Theory

There are several methods of extending homology groups to spaces other than polyhedra. Probably the most useful one is the singular homology theory, which is discussed briefly in this section. Instead of insisting that the space X be built from properly joined simplexes, one considers continuous maps from standard simplexes into X. These maps are called "singular simplexes." There are natural definitions of chains, cycles, and boundaries paralleling those of simplicial homology. In fact, the singular and simplicial theories produce isomorphic homology groups when applied to polyhedra. The singular approach, however, applies to all topological spaces, not just polyhedra.

First we define the standard simplexes which will be the domains of our singular simplexes. For notational reasons, points of $\mathbb{R}^{n+1}$ will be written $(x_0, x_1, \ldots, x_n)$ with zeroth coordinate x_0, first coordinate x_1, etc. Thus the coordinates are numbered 0 through n.

Definition. The *unit n-simplex*, $n \geq 0$, in $\mathbb{R}^{n+1}$ is the set

$$\Delta_n = \left\{(x_0, x_1, \ldots, x_n) \in \mathbb{R}^{n+1} : \sum x_i = 1, x_i \geq 0, 0 \leq i \leq n.\right\}$$

The point v_i with ith coordinate 1 and all other coordinates 0 is called the ith *vertex* of Δ_n. The subset

$$\Delta_n(i) = \{(x_0, x_1, \ldots, x_n) \in \Delta_n : x_i = 0\}$$

is called the ith *face* of Δ_n or the *face opposite the* ith *vertex*. The map $d_i : \Delta_{n-1} \to \Delta_n$ defined by

$$d_i(x_0, \ldots, x_{n-1}) = (x_0, \ldots, x_{i-1}, 0, x_i, \ldots, x_{n-1})$$

is the ith *inclusion map*.

Note that Δ_n is simply the simplex in $\mathbb{R}^{n+1}$ whose vertices are the points $v_0 = (1, 0, \ldots, 0)$, $v_1 = (0, 1, 0, \ldots, 0), \ldots, v_n = (0, \ldots, 0, 1)$. The ith inclusion map d_i maps Δ_{n-1} onto the ith face of Δ_n. For the inclusion maps in the diagram

$$\Delta_{n-2} \xrightarrow{d_j} \Delta_{n-1} \xrightarrow{d_i} \Delta_n$$
$$\Delta_{n-2} \xrightarrow{d_{i-1}} \Delta_{n-1} \xrightarrow{d_j} \Delta_n, \qquad j < i,$$

we have $d_i d_j = d_j d_{i-1}$. The proof of this is left as an exercise.

Definition. Let X be a space and n a non-negative integer. A *singular n-simplex* in X is a continuous function $s^n : \Delta_n \to X$. The set of all singular n-simplexes in X is denoted $S_n(X)$. For $n > 0$ and $0 \leq i \leq n$, the composite map

$$s_i^n = s^n d_i : \Delta_{n-1} \to X$$

is a singular $(n-1)$-simplex called the ith *face* of s^n. The function from $S_n(X)$ to $S_{n-1}(X)$ which takes a singular n-simplex to its ith face is called the ith *face operator* on $S_n(X)$. The *singular complex* of X is the set

$$S(X) = \bigcup_{n=0}^{\infty} S_n(X)$$

together with its family of face operators. It is usually denoted by $S(X)$.

Theorem 7.12. *Let s^n be a singular n-simplex in a space X, $n > 1$. Then*

$$s_{i,j}^n = s_{j,i-1}^n, \qquad 0 \leq j < i \leq n.$$

PROOF. In the notation of the preceding definitions,

$$s_{i,j}^n = s_i^n d_j = s^n d_i d_j = s^n d_j d_{i-1} = s_j^n d_{i-1} = s_{j,i-1}^n. \qquad \square$$

Definition. A *p-dimensional singular chain*, or *singular p-chain*, p a non-negative integer, is a function $c_p: S_p(X) \to \mathbb{Z}$ from the set of singular p-simplexes of X into the integers such that $c_p(s^p) = 0$ for all but finitely many singular p-simplexes. Under the pointwise operation of addition induced by the integers, the set $C_p(X)$ of all singular p-chains on X forms a group. This group is the *p-dimensional singular chain group* of X.

As in the simplicial theory, a singular p-chain can be expressed as a formal linear combination

$$c_p = \sum_{i=0}^{r} g_i \cdot s(i)^p$$

where g_i represents the value of c_p at the singular p-simplex $s(i)^p$ and c_p has value zero for all p-simplexes not appearing in the sum. Since simplicial complexes have only finitely many simplexes, the "finitely nonzero" property of p-chains holds automatically in the simplicial theory. As in the simplicial theory, algebraic systems other than the integers can be used as the set of coefficients.

Definition. The *singular boundary homomorphism*

$$\partial: C_p(X) \to C_{p-1}(X)$$

is defined for an elementary singular p-chain $g \cdot s^p$, $p \geq 1$, by

$$\partial(g \cdot s^p) = \sum_{i=0}^{p} (-1)g^i \cdot s_i^p$$

This function is extended by linearity to a homomorphism ∂ from $C_p(X)$ into $C_{p-1}(X)$. The boundary of each singular 0-chain is defined to be 0.

Theorem 7.13. *If X is a space and $p \geq 2$, then the composition $\partial\partial: C_p(X) \to C_{p-2}(X)$ in the diagram*

$$C_p(X) \xrightarrow{\partial} C_{p-1}(X) \xrightarrow{\partial} C_{p-2}(X)$$

is the trivial homomorphism.

PROOF. Since each p-chain is a linear combination of elementary p-chains, it is sufficient to prove that $\partial\partial(g \cdot s) = 0$ for each elementary p-chain $g \cdot s$. Note that

$$\begin{aligned}
\partial\partial(g \cdot s) &= \partial\left(\sum_{i=0}^{p} (-1)^i g \cdot s_i\right) = \sum_{i=0}^{p} (-1)^i \sum_{j=0}^{p-1} (-1)^j g \cdot s_{i,j} \\
&= \sum_{i=0}^{p} \sum_{j=0}^{p-1} (-1)^{i+j} g \cdot s_{i,j} \\
&= \sum_{0 \leq j < i \leq p} (-1)^{i+j} g \cdot s_{i,j} + \sum_{0 \leq i \leq j \leq p-1} (-1)^{i+j} g \cdot s_{i,j} \\
&= \sum_{0 \leq j < i \leq p} (-1)^{i+j} g \cdot s_{j,i-1} + \sum_{0 \leq i \leq j \leq p-1} (-1)^{i+j} g \cdot s_{i,j}.
\end{aligned}$$

In the left sum on the preceding line, replace $i - 1$ by j and j by i and the two sums will cancel completely. Thus $\partial\partial = 0$. □

Definition. If X is a space and p a positive integer, a *p-dimensional singular cycle* on X, or *singular p-cycle*, is a singular p-chain z_p such that $\partial(z_p) = 0$. The set of singular p-cycles is thus the kernel of the homomorphism $\partial: C_p(X) \to C_{p-1}(X)$ and is a subgroup of $C_p(X)$. This subgroup is denoted $Z_p(X)$ and called the *p-dimensional singular cycle group* of X. Since the boundary of each singular 0-chain is 0, we define singular 0-cycle to be synonymous with singular 0-chain. Then the group $Z_0(X)$ of singular 0-cycles is the group $C_0(X)$.

If $p \geq 0$, a singular p-chain b_p is a *p-dimensional singular boundary*, or *singular p-boundary*, if there is a singular $(p+1)$-chain c_{p+1} such that $\partial(c_{p+1}) = b_p$. The set $B_p(X)$ of singular p-boundaries is then the image $\partial(C_{p+1}(X))$ and is a subgroup of $C_p(X)$. This subgroup is called the *p-dimensional singular boundary group of X*. Since $\partial\partial: C_p(X) \to C_{p-2}(X)$ is the trivial homomorphism, then $B_p(X)$ is a subgroup of $Z_p(X)$, $p \geq 0$. The quotient group

$$H_p(X) = Z_p(X)/B_p(X)$$

is the *p-dimensional singular homology group* of X.

Many similarities in the definitions of the simplicial and singular homology groups should be obvious. Note, however, that no mention of orientation was made in the singular case. This was taken care of implicitly in the definition of the boundary operator:

$$\partial(g \cdot s^n) = \sum_{i=0}^{n} (-1)^i g \cdot s_i^n.$$

The definition in effect requires that the standard n-simplex Δ_n be assigned the orientation induced by the ordering $v_0 < v_1 < \cdots < v_n$. This orientation is then preserved in each singular n-simplex.

Definition. Let X and Y be spaces and $f: X \to Y$ a continuous map. If $s \in S_p(X)$, the composition fs belongs to $S_p(Y)$. Hence f induces a homomorphism

$$f_p: C_p(X) \to C_p(Y)$$

defined by

$$f_p\left(\sum_{i=0}^{r} g_i \cdot s(i)^p\right) = \sum_{i=0}^{r} g_i \cdot fs(i)^p, \qquad \sum_{i=0}^{r} g_i \cdot s(i)^p \in C_p(X).$$

One easily observes that the diagram

$$\begin{array}{ccc} C_p(X) & \xrightarrow{f_p} & C_p(Y) \\ \partial \downarrow & & \downarrow \partial \\ C_{p-1}(X) & \xrightarrow{f_{p-1}} & C_{p-1}(Y) \end{array}$$

is commutative, so f_p maps $Z_p(X)$ into $Z_p(Y)$ and $B_p(X)$ into $B_p(Y)$. (Compare with Theorem 3.1.) Thus f induces for each p a homomorphism

$$f_p^*: H_p(X) \to H_p(Y)$$

defined by

$$f_p^*(z_p + B_p(X)) = f_p(z_p) + B_p(Y), \qquad (z_p + B_p(X)) \in H_p(X).$$

The sequence $\{f_p^*\}$ is the *sequence of homomorphisms induced by f.*

The invention of singular homology theory is usually attributed to Solomon Lefschetz who introduced the singular homology groups in 1933 [45]. The basic idea can be found, however, in the classic book *Analysis Situs* [21] written by Oswald Veblen twelve years earlier. The important simplification obtained by using the ordered simplex Δ_n is due to Samuel Eilenberg.

Singular homology has two great advantages over simplicial homology: (1) The singular theory applies to all topological spaces, not just polyhedra. (2) The induced homomorphisms are defined more easily in the singular theory. Recall that in the simplicial theory a continuous map between two polyhedra must be replaced by a simplicial approximation in order to define the induced homomorphisms. This presents problems of uniqueness which are completely avoided by the singular approach. As mentioned earlier, the singular and simplicial homology groups are isomorphic for polyhedra.

The singular homology theory presented in this section is the barest of introductions. The theory has developed extensively and contains theorems paralleling those of simplicial homology. There are, for example, exact homology sequences and relative homology groups for singular homology. Anyone interested in learning more about singular homology should consult references [10] and [20].

7.5 Axioms for Homology Theory

There are homology theories other than the original simplicial theory of Poincaré and the singular theory. For example, homology groups for compact metric spaces were defined by Leopold Vietoris [56] in 1927 and for compact Hausdorff spaces by Eduard Čech [32] in 1932. The similarities of all these theories led Samuel Eilenberg (1913–) and Norman Steenrod (1910–1971) to define the general term "homology theory."

The definition applies to various categories of pairs (X, A), where X is a space with subspace A, and continuous functions on such pairs. A *homology theory* consists of three functions H, $*$, and ∂ having the following properties:

(1) H assigns to each pair (X, A) under consideration and each integer p an abelian group $H_p(X, A)$. This group is the *p-dimensional relative homology group of X modulo A*. If $A = \varnothing$ then $H_p(X, \varnothing) = H_p(X)$ is the *p-dimensional homology group* of X.

(2) If (X, A) and (Y, B) are pairs and $f\colon X \to Y$ with $f(A) \subset B$ an admissible map, then the function $*$ determines for each integer p a homomorphism

$$f_p^*\colon H_p(X, A) \to H_p(Y, B)$$

called the *homomorphism induced by f* in dimension p.

(3) The function ∂ assigns to each pair (X, A) and each integer p a homomorphism

$$\partial: H_p(X, A) \to H_{p-1}(A)$$

called the *boundary operator* on $H_p(X, A)$.

The functions H, $*$, and ∂ are required to satisfy the following seven conditions:

The Eilenberg–Steenrod Axioms

I (The Identity Axiom). *If $i: (X, A) \to (X, A)$ is the identity map, then the induced homomorphism*

$$i_p^*: H_p(X, A) \to H_p(X, A)$$

is the identity isomorphism for each integer p.

II (The Composition Axiom). *If $f: (X, A) \to (Y, B)$ and $g: (Y, B) \to (Z, C)$ are admissible maps, then*

$$(gf)_p^* = g_p^* f_p^*: H_p(X, A) \to H_p(Z, C)$$

for each integer p.

III (The Commutativity Axiom). *If $f: (X, A) \to (Y, B)$ is an admissible map and $g: A \to B$ is the restriction of f, then the diagram*

$$\begin{array}{ccc} H_p(X, A) & \xrightarrow{f_p^*} & H_p(Y, B) \\ \downarrow{\scriptstyle\partial} & & \downarrow{\scriptstyle\partial} \\ H_{p-1}(A) & \xrightarrow{g_p^*} & H_{p-1}(B) \end{array}$$

is commutative for each integer p.

IV (The Exactness Axiom). *If $i: A \to X$ and $j: (X, \varnothing) \to (X, A)$ are inclusion maps, then the homology sequence*

$$\cdots \to H_p(A) \xrightarrow{i^*} H_p(X) \xrightarrow{j^*} H_p(X, A) \xrightarrow{\partial} H_{p-1}(A) \to \cdots$$

is exact.

V (The Homotopy Axiom). *If the maps $f, g: (X, A) \to (Y, B)$ are homotopic, then the induced homomorphisms f_p^* and g_p^* are equal for each integer p.*

VI (The Excision Axiom). *If U is an open subset of X with $\overline{U} \subset A$, then the inclusion map*

$$e: (X\backslash U, A\backslash U) \to (X, A)$$

induces an isomorphism

$$e_p^*: H_p(X\backslash U, A\backslash U) \to H_p(X, A)$$

for each integer p. (The map e is called the excision *of U.)*

VII (The Dimension Axiom). *If X is a space with only one point, then*

$$H_p(X) = \{0\}$$

for each nonzero value of p.

Simplicial homology theory as presented in this book applies to the category of pairs (X, A) where X and A have triangulations K and L for which L is a subcomplex of K. The singular homology theory applies to all pairs (X, A) where X is a topological space with subspace A. For a survey of homology theory from the axiomatic point of view, see the classic book *Foundations of Algebraic Topology* by Eilenberg and Steenrod [4].

Exercises

1. Let c be a p-chain on a complex K and v a vertex for which vc is defined. Prove that

$$\partial(vc) = c - v\,\partial c.$$

2. In the proof of Theorem 7.2, show that $\psi_p(\tau^p) = \eta\cdot\sigma^p$ where η is 0, 1, or -1.

3. Show that chain homotopy is an equivalence relation for chain mappings.

4. Show that chain equivalence is an equivalence relation for complexes.

5. **Definition.** Let K be a complex and v a vertex not in K such that if $\langle v_0 \ldots v_p\rangle$ is a simplex of K, then the set $\{v, v_0, \ldots, v_p\}$ is geometrically independent. The complex vK consisting of all simplexes of K, the vertex v, and all simplexes of the form $\langle vv_0 \ldots v_p\rangle$, where $\langle v_0 \ldots v_p\rangle$ is in K, is called the *cone complex of K with respect to v.*

(a) If vK is a cone complex, prove that

$$H_0(vK) \cong \mathbb{Z}, \qquad H_p(vK) = \{0\}, \qquad p > 0.$$

(b) Show that the geometric carrier of each cone complex is contractible.

6. Complete the details in the proof of Theorem 7.5.

7. Prove the following facts about S^n:

(a) If n is even, then every continuous map on S^n of positive degree has a fixed point.

(b) If n is odd, then every continuous map on S^n of negative degree has a fixed point.

8. Prove that every continuous map from the projective plane into itself has a fixed point.

9. Let $|K|$ be a contractible polyhedron. Prove that every continuous map on $|K|$ has a fixed point.

10. Prove or disprove: If $|K|$ is a polyhedron and f, g are homotopic maps on $|K|$, then f has a fixed point if and only if g has a fixed point.

11. Give an example of a continuous map on a polyhedron that has no fixed point. Prove from the definition that the map has Lefschetz number 0.

12. Prove that $H_p(K/\varnothing) \cong H_p(K)$ for each complex K, $p \geq 0$.

13. Show that $B_p(K/L) \subset Z_p(K/L)$ for each subcomplex L of a complex K.

14. Let K be a complex and v a vertex of K. Determine the relative homology groups $H_p(K/\langle v \rangle)$, $p \geq 0$.

15. Let K be a complex of dimension n and L a subcomplex of dimension r. Prove that

$$H_p(K/L) \cong H_p(K), \qquad p \geq r + 2.$$

Is there any relation between $H_{r+1}(K/L)$ and $H_{r+1}(K)$?

16. Show that the functions i^*, j^*, and ∂^* in the homology sequence of a pair (K, L) are well-defined homomorphisms. Explain why i^* may not be one-to-one even though $i: L \to K$ is the inclusion map.

17. Prove Theorem 7.10.

18. Complete the proof of Theorem 7.11.

19. Complete the details of Example 7.7.

20. Suppose that a complex K is the union of two subcomplexes K_1 and K_2 having a single vertex in common. Determine the homology groups of K in terms of those of K_1 and K_2.

21. Show that if $j < i$, then $d_i d_j = d_j d_{i-1}$ for the inclusion maps in the diagram

$$\Delta_{n-2} \xrightarrow{d_j} \Delta_{n-1} \xrightarrow{d_i} \Delta_n$$

$$\Delta_{n-2} \xrightarrow{d_{i-1}} \Delta_{n-1} \xrightarrow{d_j} \Delta_n.$$

22. **Definition.** A subset M of a complex K is an *open subcomplex* of K means that $K \backslash M$ is a subcomplex of K.

Prove the **Excision Theorem** for simplicial homology: *Let K be a complex, L a subcomplex of K and M an open subcomplex of L. If $e: |K \backslash M| \to |K|$ is the inclusion map, then the induced homomorphism*

$$e_p^*: H_p\left(\frac{K \backslash M}{L \backslash M}\right) \to H_p\left(\frac{K}{L}\right)$$

is an isomorphism for each integer p.

23. (a) Define the term "chain mapping" for singular homology theory.
(b) Show that a continuous map $f: X \to Y$ induces a chain mapping on the associated chain groups.
(c) Define the induced homomorphisms on the singular homology groups in terms of chain mappings.

24. (a) Define the term "deformation operator" for singular homology theory.
(b) Prove that homotopic maps $f, g: X \to Y$ induce the same homomorphism

$$f_p^* = g_p^*: H_p(X) \to H_p(Y)$$

in the singular homology theory.

A Note About the Appendices

The three appendices give basic definitions and theorems about set theory, point-set topology, and algebra assumed in the text. These facts are intended to refresh the reader's memory. The appendices are not complete treatments in any sense; proofs are not included. More complete expositions and proofs for the theorems listed here can be found in many standard texts. For example, see the text by Dugundji [3] or the text by Munkres [18] for set theory and point-set topology and the text by Jacobson [12] for algebra.

APPENDIX 1
Set Theory

The symbol "$\in$" indicates set membership, and "$\subset$" indicates set inclusion. Thus $a \in A$ means that a is a member or an element of set A; $A \subset B$ means that set A is contained in set B or that A is a subset of B. The notation $\{x \in A: \ldots\}$ denotes the set of all members of A satisfying the statement...; for example, if A is the set of real members, then $\{x \in A: 0 \leq x \leq 4\}$ denotes the set of real numbers from 0 to 4 inclusive. Subsets of A other than A itself and the empty set $\varnothing$ are called *proper subsets.*

Definition. If A and B are sets, the *union* $A \cup B$ and *intersection* $A \cap B$ are defined by

$$A \cup B = \{x: x \in A \text{ or } x \in B\},$$
$$A \cap B = \{x: x \in A \text{ and } x \in B\}.$$

Unions and intersections of arbitrary families of sets are similarly defined. If $A \subset X$, then the *complement* of A with respect to X is the set $X \backslash A$ of members of X which do not belong to A:

$$X \backslash A = \{x \in X: x \notin A\}.$$

Definition. The *Cartesian product* of two sets A and B is the set

$$A \times B = \{(a, b): a \in A \text{ and } b \in B\}.$$

The Cartesian product of a finite collection $\{A_i\}_{i=1}^n$, where each A_i is a set, is defined analogously:

$$A_1 \times A_2 \times \cdots \times A_n = \{(a_1, a_2, \ldots, a_n): a_i \in A_i, 1 \leq i \leq n\}.$$

The point a_i is called the ith *coordinate* of $(a_1, a_2, \ldots, a_n)$.

Products can be defined for arbitrary families of sets; this must be postponed, however, until the concept of function from one set to another has been introduced.

Definition. A *relation* from set A to set B is a subset $\sim$ of the Cartesian product $A \times B$. It is customary and simpler to write $a \sim b$ to mean $(a, b) \in \sim$. A relation $\sim$ from A to itself is an *equivalence relation* means that the following three properties are satisfied:

(1) *The Reflexive Property:* $x \sim x$ for all $x \in A$.
(2) *The Symmetric Property:* If $x \sim y$ then $y \sim x$.
(3) *The Transitive Property:* If $x \sim y$ and $y \sim z$, then $x \sim z$.

The *equivalence class* of x is the set

$$[x] = \{y \in A: x \sim y\}.$$

If $\sim$ is an equivalence relation on A, then each element of A belongs to exactly one equivalence class.

Definition. A *function* $f: A \to B$ is a relation from set A to set B such that if $a \in A$ there is only one $b \in B$ for which afb. It is customary to write $f(a) = b$ and to call b the *image* of a under f. Set A is the *domain* of f, and the *range* of f is the set

$$f(A) = \{b \in B: b = f(a) \text{ for some } a \in A\}.$$

Definition. If $f: A \to B$ and $g: B \to C$ are functions on the indicated sets, then the *composite function* $gf: A \to C$ is defined by

$$gf(a) = g(f(a)), \qquad a \in A.$$

Definition. The identity function on a set A is the function $i: A \to A$ such that $i(a) = a$ for all $a \in A$.

Definition. A function $f: A \to B$ is *one-to-one* if no two members of A have the same image; f is *onto* if $f(A) = B$. A function which is both one-to-one and onto is called a *one-to-one correspondence*. Thus a one-to-one correspondence is a function from A to B for which each point of B is the image of exactly one point of A. In this case there is an *inverse function* $f^{-1}: B \to A$ defined by: $a = f^{-1}(b)$ if and only if $b = f(a)$.

If $f: A \to B$ is a one-to-one correspondence, then the composite functions $f^{-1}f$ and ff^{-1} are the identity functions on A and B respectively.

Definition. If there is a one-to-one correspondence between sets A and B, then A and B are said to have the *same cardinal number*.

Definition. If $f\colon A \to B$ is a function and $C \subset A$, the *restriction* $f|_C\colon C \to B$ of f to C is the function with domain C defined by

$$f|_C(x) = f(x), \qquad x \in C.$$

Equivalently, f is called an *extension* of $f|_C$.

Definition. If $\{A_j\}$ is a family of sets indexed by a set J (i.e., if A_j is a set for each j in a given set J), then the *product* of the sets A_j is the set $\prod_{j\in J} A_j$ composed of all functions $f\colon J \to \bigcup A_j$ such that $f(j) \in A_j$ for each $j \in J$.

The finite product $A_1 \times A_2 \times \cdots \times A_n$ is a special case of the preceding definition. Let J be the set of integers $1, 2, \ldots, n$, and identify the sequence $(a_1, a_2, \ldots, a_n)$ with the function $f\colon J \to \bigcup_{j=1}^{n} A_j$ whose value at j is a_j. Then

$$A_1 \times A_2 \times \cdots \times A_n = \prod_{j\in J} A_j.$$

Definition. Let $f\colon X \times Y \to Z$ be a function from the product set $X \times Y$ into Z. If x_0 is a point of X, then the symbol $f(x_0, \cdot)$ denotes the function from Y into Z defined by

$$f(x_0, \cdot)(y) = f(x_0, y), \qquad y \in Y.$$

For y_0 in Y, the symbol $f(\cdot, y_0)$ denotes the function from X to Z defined by

$$f(\cdot, y_0)(x) = f(x, y_0), \qquad x \in X.$$

APPENDIX 2

Point-set Topology

Definition. A *topology* for a set X is a family T of subsets of X satisfying the following three properties:

(1) The set X and the empty set $\varnothing$ are in T.
(2) The union of any family of members of T is in T.
(3) The intersection of any finite family of members of T is in T.

The members of T are called *open sets.* A *topological space*, or simply *space*, is a pair (X, T) where X is a set and T is a topology for X. One often refers to a topological space X, omitting mention of the topology, when the name of the topology is not important.

A *base* or *basis* for a topology T is a subset B of T such that each member of T is a union of members of B. *A subbase* or *subbasis* for T is a subset S of T such that the family of all finite intersections of members of S is a basis for T.

If X is a space, a subset C of X is *closed* means that its complement $X \backslash C = \{x \in X : x \notin C\}$ is open. A *neighborhood* of a point x in X is an open set containing x.

A point x is a *limit point* of a subset A of X means that every neighborhood of x contains a point of A distinct from x. The *closure* of a set A is the set $\overline{A}$, the union of A with its set of limit points. The *boundary* of A is the intersection of $\overline{A}$ with $\overline{X \backslash A}$.

Proposition. *A subset A of a space X is closed if and only if A contains all its limit points. A subset O of X is open if and only if O contains a neighborhood of each of its points. The closure of each subset of X is a closed set.*

Definition. A space X is a *Hausdorff space* or a *T_2-space* provided that for each pair x_1, x_2 of distinct points of X there exist disjoint neighborhoods O_1 and O_2 of x_1 and x_2 respectively.

Definition. The *subspace topology* for a subset A of a space X consists of all subsets of the form $O \cap A$ where O is open in X. The set A with its subspace topology is a *subspace* of X.

Definition. A *covering* $\mathscr{C}$ of a space X is a family of subsets of X whose union is X. A *subcovering* of $\mathscr{C}$ is a covering each of whose members is a member of $\mathscr{C}$. A covering each of whose members is an open set is called an *open covering*.

Definition. A space X is *compact* provided that every open covering of X has a finite subcovering. A *compact subset* of X is a subset which is compact in its subspace topology. A space is *locally compact* means that for each point x there is a neighborhood U of x and a compact set A with $U \subset A$.

Proposition. (a) *In a Hausdorff space, compact sets are closed.*

(b) *A closed subset of a compact space is compact.*

(c) *If X is a locally compact Hausdorff space and $x \in X$, then for each neighborhood V of x there is a neighborhood O of x such that $\bar{O} \subset V$ and $\bar{O}$ is compact.*

Definition. A space X is *connected* means that X is not the union of two disjoint, nonempty open sets. A *connected subset* of X is a subset which is connected in its subspace topology. A *component* is a connected subset which is not a proper subset of any connected subset of X.

Definition. A *metric* or *distance function* for a set X is a function d from the Cartesian product $X \times X$ to the non-negative real numbers such that, for all x, y, z in X,

(1) $d(x, y) = d(y, x)$,
(2) $d(x, y) = 0$ if and only if $x = y$,
(3) $d(x, y) + d(y, z) \geq d(x, z)$.

For $x \in X$ and $r > 0$ the set

$$S(x, r) = \{y \in X: d(x, y) < r\}$$

is called the *spherical neighborhood with center x and radius r*. The set of all such spherical neighborhoods is a basis for a topology for X, the *metric topology* determined by d. A set with the topology determined by a metric is called a *metric space*. The *diameter* of a subset A of a metric space is the least upper bound of the distances between points of A:

$$\operatorname{diam} A = \operatorname{lub}\{d(x, y): x, y \in A\}.$$

A set with finite diameter is called *bounded*.

Definition. A function $f: X \to Y$ from a space X to a space Y is *continuous* provided that for each open set U in Y the inverse image

$$f^{-1}(U) = \{x \in X: f(x) \in U\}$$

is open in X. A one-to-one correspondence $f: X \to Y$ for which both f and the inverse function f^{-1} are continuous is called a *homeomorphism*; in this case X and Y are said to be *homeomorphic*. A function $g: X \to Y$ is *open* provided that $g(O)$ is open in Y for each open subset O of X. *Closed function* is defined analogously.

Proposition. *The composition of continuous functions is continuous.*

Proposition. *The properties of being compact or connected are preserved by continuous functions.*

Proposition. *Let $f: X \to Y$ be a function on the indicated spaces. The following statements are equivalent:*

(a) *f is continuous.*
(b) *For each closed subset C of Y, $f^{-1}(C)$ is closed in X.*
(c) *There is a basis B for Y such that $f^{-1}(U)$ is open for each $U \in B$.*
(d) *There is a subbasis S for Y such that $f^{-1}(U)$ is open for each $U \in S$.*

Proposition. *If X and Y are metric spaces with metrics d and d' respectively and $f: X \to Y$ is a function, then f is continuous if and only if for each $x_0 \in X$ and $\epsilon > 0$, there is a number $\delta > 0$ such that if $d(x_0, x) < \delta$, then $d(f(x_0), f(x)) < \epsilon$.*

Definition. Let X and Y be metric spaces with metrics d, d' respectively. A function $f: X \to Y$ is *uniformly continuous* means that for each $\epsilon > 0$, there is a number $\delta > 0$ such that if x and x' are points of X with $d(x, x') < \delta$, then $d(f(x), f(x')) < \epsilon$.

Proposition. *If X and Y are metric spaces, X is compact, and $f: X \to Y$ is continuous, then f is uniformly continuous.*

Proposition. *If $\mathscr{U}$ is an open covering of a compact metric space X, then there is a positive number η such that each subset of X of diameter less than η is contained in a member of $\mathscr{U}$. (The number η is called a* Lebesgue number *for the open covering $\mathscr{U}$.)*

Definition. Let X and Y be spaces. The *product space* $X \times Y$ is the Cartesian product of X and Y with the *product topology* which has as a basis the family of all sets of the form $U_1 \times U_2$ where U_1 is open in X and U_2 is open in Y.

If $\{X_\alpha\}$ is a family of spaces indexed by a set A, then the *product space* $\prod_{\alpha \in A} X_\alpha$ is the product of the sets X_α with the *product topology* which has as a subbasis all sets of the form $p_\beta^{-1}(U_\beta)$, $\beta \in A$. Here $p_\beta\colon \prod_{\alpha \in A} X_\alpha \to X_\beta$ is the *projection* on X_β defined by

$$p_\beta(f) = f(\beta), \qquad f \in \prod_{\alpha \in A} X,$$

and U_β represents an arbitrary open set in X_β.

Proposition. (a) *A product of compact spaces is compact.*

(b) *A product of connected spaces is connected.*

(c) *If $x_0 \in X$ and $y_0 \in Y$, then the subspaces $X \times \{y_0\}$ and $\{x_0\} \times Y$ of $X \times Y$ are homeomorphic to X and Y respectively.*

Definition. Let X be a space and S an equivalence relation on X. Then S partitions X into a family X/S of equivalence classes. The *quotient topology* for X/S is defined by the following condition: A set U of equivalence classes in X/S is open if and only if the union of the members of U is open in X. The *quotient space of X modulo S* is the set X/S with the quotient topology.

As an important special case we have the quotient space X/A where A is a subset of X. This is the quotient space of X determined by the relation: xSy if and only if $x = y$ or x and y are both in A. The points of X/A are the points of $X\backslash A$ and an additional single point A.

If $f\colon X \to Y$ is a function from a space X onto a set Y, then the *quotient topology* for Y consists of all sets $U \subset Y$ for which $f^{-1}(U)$ is open in X. The function f determines an equivalence relation R on X defined by $x_1 R x_2$ if and only if $f(x_1) = f(x_2)$. The quotient space X/R is homeomorphic to the space Y with the quotient topology determined by f.

Proposition. *Let $f\colon X \to Y$ be a continuous function from space X onto space Y. If f is either open or closed, then Y has the quotient topology determined by f.*

Definition. *Euclidean n-dimensional space* $\mathbb{R}^n$, n a positive integer, is the set

$$\mathbb{R}^n = \{x = (x_1, \ldots, x_n)\colon x_i \text{ is a real number, } 1 \le i \le n\}$$

with the topology determined by the *Euclidean metric*:

$$d(x, y) = \left\{\sum_{i=1}^{n} (x_i - y_i)^2\right\}^{1/2}$$

where $x = (x_1, \ldots, x_n)$ and $y = (y_1, \ldots, y_n)$ are members of $\mathbb{R}^n$. The members of $\mathbb{R}^n$ are referred to as *points* or *vectors*. The *norm* or *length* $\|x\|$ of a vector x in $\mathbb{R}^n$ is the distance from x to the origin $0 = (0, \ldots, 0)$:

$$\|x\| = \left\{\sum_{i=1}^{n} x_i^2\right\}^{1/2}.$$

Note that $\mathbb{R}^1$ is simply the real number line: $\mathbb{R}^1 = \mathbb{R}$.

For x, y in $\mathbb{R}^n$, the *inner product* or *dot product* of x and y is the number

$$x \cdot y = x_1y_1 + x_2y_2 + \cdots + x_ny_n.$$

The vectors x and y are *perpendicular* or *orthogonal* if $x \cdot y = 0$. This definition extends the common concept of perpendicularity in two and three dimensions to higher dimensions.

The *unit n-sphere* S^n is the set of all points in $\mathbb{R}^{n+1}$ of unit length:

$$S^n = \{x = (x_1, \ldots, x_{n+1}) \in \mathbb{R}^{n+1}\colon \|x\| = 1\}, \qquad n \geq 0.$$

Note that S^n is a subspace of $\mathbb{R}^{n+1}$, not of $\mathbb{R}^n$. We may consider $\mathbb{R}^n$ as the subspace of $\mathbb{R}^{n+1}$ consisting of all points having final coordinate 0.

Proposition. (a) *Euclidean n-space is homeomorphic to the product of n copies of the space of real numbers.*

(b) *A subspace of $\mathbb{R}^n$ is compact if and only if it is closed and bounded.*

Definition. The *unit n-ball* B^n is the set of all points in $\mathbb{R}^n$ of length not exceeding 1:

$$B^n = \{x = (x_1, \ldots, x_n) \in \mathbb{R}^n\colon \|x\| \leq 1\}, \qquad n \geq 1.$$

Note that the boundary of B^n is the unit $(n-1)$-sphere S^{n-1}.

The *unit n-cube* I^n is the set

$$I^n = \{t = (t_1, \ldots, t_n) \in \mathbb{R}^n\colon 0 \leq t_i \leq 1 \text{ for each } i\}.$$

Thus $I^1 = I$ is the closed unit interval $[0, 1]$, I^2 is a square, and I^3 is a 3-dimensional cube. The *boundary* of I^n, denoted ∂I^n, is the set of all points of I^n having some coordinate equal to 0 or 1.

Proposition. (a) *The quotient space of B^n obtained by identifying its boundary S^{n-1} to a single point is homeomorphic to S^n.*

(b) *The quotient space of I^n obtained by identifying its boundary ∂I^n to a single point is homeomorphic to S^n.*

Definition. Let X be a Hausdorff space which is not compact and ∞ a point not in X. The *one-point compactification* X^* of X is the set

$$X^* = X \cup \{\infty\}$$

with the topology determined by the basis composed of all open sets in X together with all subsets U of X^* for which $X^* \backslash U$ is a closed, compact subset of X.

Proposition. *The one-point compactification X^* of a Hausdorff space X is a compact space; X^* is Hausdorff if and only if X is locally compact.*

Proposition. *The one-point compactification of Euclidean n-space $\mathbb{R}^n$ is homeomorphic to S^n.*

APPENDIX 3

Algebra

Definition. A *binary operation* on a set A is a function $f: A \times A \to A$. For $a, b \in A$, $f(a, b)$ is often expressed ab or $a \cdot b$ (multiplicative notation) or $a + b$ (additive notation).

Definition. A *group* is a set G together with a binary operation on G satisfying the following three properties:

(1) $a(bc) = (ab)c$ for all $a, b, c \in G$.
(2) There is an element e, the *identity element* of G, such that $ae = ea = a$ for all a in G.
(3) For each a in G, there is an element a^{-1}, the *inverse* of a, such that $aa^{-1} = a^{-1}a = e$.

In the additive group notation, the identity element is denoted by 0 and the inverse of a by $-a$. A group whose only element is the identity is the *trivial group* $\{0\}$.

A subset A of a group G is a *subgroup* of G provided that A is a group under the operation of G. If A is a subgroup and $g \in G$, then

$$gA = \{ga: a \in A\}$$

is called the *left coset of A by g*. In the additive notation, we would write $g + A$ instead of gA. *Right cosets* are defined similarly.

Proposition. *Left cosets gA and hA of a subgroup A are either disjoint or identical.*

Definition. A group G is *commutative* or *abelian* means that $ab = ba$ for all $a, b \in G$.

Definition. A *homomorphism* $f: G \to H$ from a group G into a group H is a function such that

$$f(ab) = f(a)f(b), \qquad a, b \in G.$$

The set

$$\operatorname{Ker} f = \{a \in G: f(a) = \text{identity of } H\}$$

is the *kernel* of f. An *isomorphism* is a homomorphism which is also a one-to-one correspondence between G and H; in this case the groups are called *isomorphic*, and we write $G \simeq H$.

Definition. A subgroup A of a group G is *normal* means that $g^{-1}ag \in A$ for all $g \in G$, $a \in A$.

Proposition. *The kernel of a homomorphism $f: G \to H$ is a normal subgroup of G. The homomorphism is one-to-one if and only if the kernel of f contains only the identity of G.*

Proposition. *If A is a normal subgroup of G, then each left coset gA equals the corresponding right coset Ag. The family G/A of all left cosets of A is a group under the operation*

$$gA \cdot hA = ghA.$$

(*The group G/A is called the* quotient group of G modulo A.)

Proposition (The First Homomorphism Theorem). *Let $f: G \to H$ be a homomorphism from group G onto group H with kernel K. Then H is isomorphic to the quotient group G/K.*

Definition. A *commutator* in a group G is an element of the form $aba^{-1}b^{-1}$. The *commutator subgroup* of G is the smallest subgroup containing all commutators of G. Equivalently, the commutator subgroup consists of all finite products of commutators of G.

Proposition. (a) *The commutator subgroup F of a group G is normal.*

(b) *The commutator subgroup is the smallest subgroup of G for which G/F is abelian.*

Definition. If g is a member of a group G, the set of all powers $g, g^{-1}, gg = g^2, g^{-1}g^{-1} = g^{-2}, \ldots$ forms a subgroup

$$[g] = \{g^n: n \text{ is an integer}\}$$

called the *subgroup generated by* g. If G has an element g for which $[g] = G$, then G is a *cyclic group* with *generator* g.

The most common cyclic group is the group $\mathbb{Z}$ of integers. Both 1 and -1 are generators.

Definition. A *set of generators* for a group G is a subset S of G such that each member of G is a product of powers of members of S. A group which has a finite set of generators is called *finitely generated.*

Definition. The *direct sum* $G \oplus H$ of groups G and H is the set $G \times H$ with operation $\oplus$ defined by

$$(g_1, h_1) \oplus (g_2, h_2) = (g_1 + g_2, h_1 + h_2)$$

for all $g_1, g_2 \in G$, $h_1, h_2 \in H$. (Here we are using additive notation.)

Definition. A group which is isomorphic to a finite direct sum of copies of the group $\mathbb{Z}$ of integers is called a *free abelian group*. Thus a free abelian group on n generators is isomorphic to the direct sum $\mathbb{Z} \oplus \mathbb{Z} \oplus \cdots \oplus \mathbb{Z}$ (n summands). The integer n is called the *rank* of the group.

Proposition. *Every subgroup of a free abelian group is a free abelian group.*

Proposition (The Decomposition Theorem for Finitely Generated Abelian Groups). *Each finitely generated abelian group is a direct sum of a free abelian group G and a finite subgroup. The finite subgroup* (*called the* torsion subgroup) *is composed of the identity element alone or is a direct sum of cyclic groups of prime power orders. The rank of G and the orders of the cyclic subgroups* (*with their multiplicities*) *are uniquely determined.*

Definition. A *permutation* on a finite set V is a one-to-one function from V onto itself. The set of all permutations on a set of n distinct objects forms a group, the *symmetric group on n objects*, under the operation of composition. A *transposition* on V is a permutation which interchanges precisely two members of V and acts as the identity map for the other members.

Proposition. *Every permutation is a product of transpositions.*

If a permutation is the product of an even number of transpositions, then it is called an *even permutation.* Although it is not obvious, it is true that if a given permutation can be represented as a product of an even number of transpositions, then every representation of it as a product of transpositions requires an even number. A permutation which is not even is called an *odd permutation.*

Example. To illustrate the way even and odd permutations are used in the text, consider a set $V = \{v_1, v_2, v_3\}$ of three elements with a definite order v_1, v_2, v_3. The arrangement v_1, v_3, v_2 represents an odd permutation of the given order since it was produced by transposing one pair of elements. Likewise, the ordering v_2, v_1, v_3 represents an odd permutation. On the other hand, v_2, v_3, v_1 represents an even permutation since it is produced from the original order by two transpositions: beginning with v_1, v_2, v_3 transpose v_1 and v_2 to produce v_2, v_1, v_3; now transpose v_1 and v_3 to produce v_2, v_3, v_1.

Definition. A *topological group* is a group G with a topology under which the operation of G is a continuous map from $G \times G$ to G and the function $g \to g^{-1}$ is a homeomorphism from G onto G.

Definition. A *ring* is a triple $(R, +, \cdot)$, where R is a set with operations $+$ and $\cdot$ (indicated by juxtaposition), such that

(1) $(R, +)$ is an abelian group,
(2) $(ab)c = a(bc)$,
(3) $a(b + c) = ab + ac$,
(4) $(b + c)a = ba + ca, \qquad a, b, c \in R$.

The operation $+$ is called *addition*, and $\cdot$ is called *multiplication*. The additive identity element is denoted by 0. If there is an identity element 1 for multiplication, then R is a *ring with unity*. A ring is *commutative* if $ab = ba$ for all $a, b \in R$.

Definition. A *field* is a commutative ring with unity in which the nonzero elements form a group under multiplication.

The most common fields are the real numbers, the rational numbers, and the complex numbers.

Definition. A *vector space* over a field F is a set V with two operations, an addition $+$ under which V forms an abelian group, and *scalar multiplication* which associates with each $v \in V$ and $a \in F$ a member av in V. The following conditions must be satisfied for all $a, b \in F$ and all $u, v \in V$:

(1) $(ab)v = a(bv)$,
(2) $a(u + v) = au + av$, $(a + b)v = av + bv$,
(3) $1 \cdot v = v$.

The members of a vector space V are called *vectors*.

Definition. A set $\{v_1, \ldots, v_k\}$ of members of a vector space V is *linearly dependent* if there exist elements $a_1, \ldots, a_k$ of the field F such that

$$a_1v_1 + \cdots + a_kv_k = 0$$

and not all the a_i are 0. A set of vectors is *linearly independent* if it is not linearly dependent. A set of vectors $\{v_1, \ldots, v_k\}$ is said to *span* V if each element $v \in V$ can be represented as a *linear combination*

$$v = b_1v_1 + \cdots + b_kv_k$$

for some $b_1, \ldots, b_k$ in F. A *base* or *basis* for V is a linearly independent set which spans V. If V has a finite basis, then V is called *finite dimensional.*

Proposition. *Any two bases for a finite dimensional vector space V have the same number of elements. (This number is the* dimension *of V.)*

The most common vector spaces are the Euclidean spaces $\mathbb{R}^n$ over the field of real numbers. Vector addition and scalar multiplication are defined by

$$(x_1, \ldots, x_n) + (y_1, \ldots, y_n) = (x_1 + y_1, \ldots, x_n + y_n),$$
$$a(x_1, \ldots, x_n) = (ax_1, \ldots, ax_n).$$

It is sometimes said that these operations are defined "componentwise" by addition and multiplication of real numbers. The vector space dimension of $\mathbb{R}^n$ is n.

Definition. A *subspace* A of a vector space V is a subset of V which is a vector space under the addition and scalar multiplication of V. A *hyperplane* is a translation of a subspace: H is a hyperplane provided that there is a subspace A and a vector $v \in V$ such that

$$H = \{v + a: a \in A\}.$$

Definition. The *sum* $A + B$ of subspaces A and B of a vector space V is the subspace

$$A + B = \{a + b: a \in A, b \in B\}.$$

If each element v in $A + B$ has a unique representation $v = a + b$ for $a \in A$ and $b \in B$, then $A + B$ is written $A \oplus B$ and called a *direct sum.*

Proposition. (a) *The sum $A + B$ is a direct sum if and only if $A \cap B = \{0\}$.*

(b) *If $A \cap B = \{0\}$ and $\{v_1, \ldots, v_k\}$ and $\{w_1, \ldots, w_j\}$ are bases for A and B respectively, then $\{v_1, \ldots, v_k, w_1, \ldots, w_j\}$ is a basis for $A \oplus B$. In particular, the dimension of $A \oplus B$ is the sum of the dimensions of A and B.*

Definition. If V and W are vector spaces over a common field F, a function $f: V \to W$ satisfying

$$f(u + v) = f(u) + f(v),$$
$$f(au) = af(u), \qquad a \in F,\ u, v \in V,$$

is called a *homomorphism* or a *linear transformation.* A one-to-one linear transformation from V onto W is an *isomorphism.*

Definition. If m and n are positive integers, an $m \times n$ *matrix* over a field F is a rectangular array

$$A = (a_{ij}) = \begin{bmatrix} a_{11}a_{12} & \cdots & a_{1n} \\ a_{21}a_{22} & \cdots & a_{2n} \\ \vdots & & \vdots \\ a_{m1}a_{m2} & \cdots & a_{mn} \end{bmatrix}$$

of mn members of F. The element a_{ij} in row i and column j is called the (i, j)th *component* of A. If $B = (b_{ij})$ is another $m \times n$ matrix, then the *matrix sum* $A + B$ is defined by

$$A + B = (a_{ij} + b_{ij}).$$

The *matrix product* AC is defined for any matrix $C = (c_{kj})$ of n rows by

$$AC = (d_{ij})$$

where $d_{ij} = \sum_{k=1}^{n} a_{ik}c_{kj}$. The elements $e_{11}, e_{22}, \ldots, e_{nn}$ of an $n \times n$ matrix $E = (e_{ij})$ are called its *diagonal elements.* The *trace* of E is the sum of its diagonal elements:

$$\text{trace } E = \sum_{i=1}^{n} e_{ii}.$$

Proposition. *Let V be a finite dimensional vector space over F with basis $\{v_1, \ldots, v_n\}$. Then there is a one-to-one correspondence between the set of linear transformations $f: V \to V$ and the set of $n \times n$ matrices over F. The matrix corresponding to f is the matrix $A_f = (a_{ij})$ where*

$$f(v_i) = \sum_{j=1}^{n} a_{ij}v_j.$$

The composition of two linear maps corresponds to the product of their associated matrices.

Proposition. *Let $f: V \to V$ be a linear transformation. If matrices B and C represent f relative to different bases, then B and C have the same trace.*

Definition. Let F be a field, and let V_n denote the vector space of all n-tuples of members of F with operations defined by

$$(x_1, \ldots, x_n) + (y_1, \ldots, y_n) = (x_1 + y_1, \ldots, x_n + y_n),$$
$$a(x_1, \ldots, x_n) = (ax_1, \ldots, ax_n).$$

If $A = (a_{ij})$ is an $m \times n$ matrix over F, then each row

$$a_{i1}a_{i2}\cdots a_{in}$$

of A can be considered a member

$$(a_{i1}, a_{i2}, \ldots, a_{in})$$

of V_n. In this context, the rows of A are called *row vectors.* The *rank* of A, rank(A), is the dimension of the subspace of V_n spanned by the row vectors of A.

Bibliography

Books

1. Ahlfors, L. V. and Sario, L. *Riemann Surfaces*, Princeton University Press, Princeton, N.J., 1960.
2. Cairns, S. S. *Introductory Topology*, Ronald Press Company, New York, 1961.
3. Dugundji, J. *Topology*, Allyn and Bacon, Inc., Boston, 1966.
4. Eilenberg, S. and Steenrod, N. *Foundations of Algebraic Topology*, Princeton University Press, Princeton, N.J., 1952.
5. Gray, B. *Homotopy Theory*, Academic Press, New York, 1975.
6. Greenberg, M. J. *Lectures on Algebraic Topology*, W. A. Benjamin, Inc., New York, 1967.
7. Hausdorff, F. *Grundzüge der Mengenlehre* (Second Edition), Walter de Gruyter, Leipzig, 1914.
8. Hilton, P. J. *An Introduction to Homotopy Theory*, Cambridge University Press, Cambridge, 1953.
9. Hocking, J. G. and Young, G. S. *Topology*, Addison-Wesley Publishing Co., Inc., Reading, Mass., 1961.
10. Hu, S. T. *Homology Theory*, Holden-Day, Inc., San Francisco, 1966.
11. Hu, S. T. *Homotopy Theory*, Academic Press, Inc., New York, 1959.
12. Jacobson, N. *Basic Algebra, I*, W. H. Freeman and Co., San Francisco, 1974.
13. Lefschetz, S. *Algebraic Topology*, American Mathematical Society Colloquium Publications, Volume XXVII, 1942.
14. Lefschetz, S. *Introduction to Topology*, Princeton University Press, Princeton, N.J., 1949.
15. Lefschetz, S. *Topology* (Second Edition), Chelsea Publishing Company, New York, 1956.
16. Massey, W. S. *Algebraic Topology: An Introduction*, Harcourt, Brace and World, Inc., New York, 1967, Springer-Verlag, 1977.

17. Maunder, C. R. F. *Algebraic Topology*, Van Nostrand Reinhold Co., London, 1970.

18. Munkres, J. R. *Topology: A First Course*, Prentice-Hall, Inc., Englewood Cliffs, New Jersey, 1975.

19. Seifert, H. and Threlfall, W. *Lehrbuch der Topologie*, Chelsea Publishing Co., New York, 1945.

20. Spanier, E. H. *Algebraic Topology*, McGraw-Hill Book Co., New York, 1966.

21. Veblen, O. *Analysis Situs* (Second Edition), American Mathematical Society Colloquium Publications, Volume V, part II, 1931.

22. Wall, C. T. C. *A Geometric Introduction to Topology*, Addison-Wesley Publishing Co., Reading, Mass., 1972.

23. Weyl, H. *The Concept of a Riemann Surface* (Third Edition), Addison-Wesley Publishing Co., Reading, Mass., 1964.

24. Whitehead, G. W. *Homotopy Theory*, M.I.T. Press, Cambridge, Mass., 1966.

Papers

25. Alexander, J. W. "A proof of the invariance of certain constants of analysis situs," *Trans. Am. Math. Soc.* **16**, 148–154 (1915).

26. Alexander, J. W. "Note on two three-dimensional manifolds with the same group," *Trans. Am. Math. Soc.* **20**, 339–342 (1919).

27. Alexander, J. W. "Combinatorial analysis situs," *Trans. Am. Math. Soc.* **28**, 301–329 (1926).

28. Borsuk, K. "Drei Sätze über die *n*-dimensionale euklidische Sphäre," *Fundamenta Math.* **20**, 177–190 (1933).

29. Brouwer, L. E. J. "On continuous vector distributions on surfaces," *Proc. Akad. von Wetenschappen, Amsterdam*, **11**, 850–858 (1909); **12**, 716–734 (1910); **13**, 171–186 (1910).

30. Brouwer, L. E. J. "Zur analysis situs," *Math. Annalen* **68**, 422–434 (1910).

31. Brouwer, L. E. J. "Beweis der Invarianz der Dimensionenzahl," *Math. Annalen* **70**, 161–165 (1911).

32. Čech, E. "Théorie générale de l'homologie dans un espace quelconque," *Fundamenta Math.* **19**, 149–183 (1932).

33. Čech, E. "Höherdimensionale Homotopiegruppen," *Proc. Int. Congr. Math.* **3**, 203 (Zürich, 1932).

34. Eilenberg, S. "Singular homology theory," *Ann. of Math.* **45**, 407–447 (1944).

35. Eilenberg, S. and Steenrod, N. "Axiomatic approach to homology theory," *Proc. Nat. Acad. Sci. U.S.A.* **31**, 117–120 (1945).

36. Fox, R. H. "On topologies for function spaces," *Bull. Am. Math. Soc.* **51**, 429–432 (1945).

37. Freudenthal, H. "Über die Klassen von Sphärenabbildungen," *Compositio Math.* **5**, 299–314 (1937).

38. Hopf, H. "Über die Abbildungen des 3-Sphäre auf die Kugelfläche," *Math. Annalen* **104**, 637–655 (1931).

39. Hopf, H. "Über die Abbildungen von Sphären auf Sphären niedrigerer Dimension," *Fundamenta Math.* **25**, 427–440 (1935).

40. Hopf, H. "A new proof of the Lefschetz formula on invariant points," *Proc. Nat. Acad. Sci. U.S.A.* **14**, 149–153 (1928).

41. Hopf, H. "Die Klassen der Abbildungen der n-dimensionalen Polyder auf die n-dimensionale Sphäre," *Comment. Math. Helvetici* **5**, 39–54 (1933).

42. Hurewicz, W. "Beiträge zur Topologie der Deformationen." I, "Höherdimensionale Homotopiegruppen," *Proc. Acad. von Wetenschappen, Amsterdam* **38**, 112–119 (1935); II, "Homotopie- und Homologiegruppen," **38**, 521–528 (1935); III, "Klassen und Homologiegruppen von Abbildungen," **39**, 117–126 (1936); IV, "Asphärische Räume," **39**, 215–224 (1936).

43. Hurewicz, W. "On duality theorems," *Bull. Am. Math. Soc.* **47**, 562–563 (1941).

44. Kirby, R. and Siebenmann, L. "On the triangulation of manifolds and the Hauptvermutung," *Bull. Am. Math. Soc.* **75**, 742–749 (1969).

45. Lefschetz, S. "On singular chains and cycles," *Bull. Am. Math. Soc.* **39**, 124–129 (1933).

46. Lefschetz, S. "The residual set of a complex on a manifold and related questions," *Proc. Nat. Acad. Sci. U.S.A.* **13**, 614–622 (1927).

47. Lefschetz, S. "Intersections and transformations of complexes and manifolds," *Trans. Am. Math. Soc.* **28**, 1–49 (1926).

48. Lefschetz, S. "Manifolds with a boundary and their transformations," *Trans. Am. Math. Soc.* **29**, 429–462 (1927).

49. Poincaré, H. "Analysis situs," *Jour. École Polytech.* **(2)1**, 1–121 (1895).

50. Poincaré, H. "Complément à l'analysis situs," *Rend. Circ. Mat., Palermo* **13**, 285–343 (1899).

51. Poincaré, H. "Deuxième complément à l'analysis situs," *Proc. London Math. Soc.* **32**, 277–308 (1900).

52. Poincaré, H. "Cinquième complément à l'analysis situs," *Rend. Circ. Mat., Palermo* **18**, 45–110 (1904).

53. Poincaré, H. "Sur un théorème de la théorie générale des fonctions," *Bull. Soc. Math. France* **11**, 112–125 (1883).
(The above papers by Poincaré can also be found in *Œuvres de Henri Poincaré*, Gauthier-Villars, Paris, 1950.)

54. Smale, S. "Generalized Poincaré's Conjecture in dimensions greater than 4," *Ann. Math.* **74**, 391–406 (1961).

55. Veblen, O. "Theory of plane curves in non-metrical analysis situs," *Trans. Am. Math. Soc.* **6**, 83–98 (1905).

56. Vietoris, L. "Über den höheren Zusammenhang kompakten Räume und eine Klasse von zusammenhangstreuen Abbildungen," *Math. Annalen* **97**, 454–472 (1927).

57. Whitehead, J. H. C. "Combinatorial homotopy, I," *Bull. Am. Math. Soc.* **55**, 213–245 (1949).

58. Yoneyama, K. "Theory of continuous sets of points," *Tohoku Math. Jour.* **11**, 43–158 (1917).

Index

Algebraic Topology: An Introduction

by **W. S. Massey**
(Graduate Texts in Mathematics, Vol. 56)
1977. xxi, 261p. 61 illus. cloth

Here is a lucid examination of algebraic topology, designed to introduce advanced undergraduate or beginning graduate students to the subject as painlessly as possible. *Algebraic Topology: An Introduction* is the first textbook to offer a straight-forward treatment of "standard" topics such as 2-dimensional manifolds, the fundamental group, and covering spaces. The author's exposition of these topics is stripped of unnecessary definitions and terminology and complemented by a wealth of examples and exercises.

Algebraic Topology: An Introduction evolved from lectures given at Yale University to graduate and undergraduate students over a period of several years. The author has incorporated the questions, criticisms and suggestions of his students in developing the text. The prerequisites for its study are minimal: some group theory, such as that normally contained in an undergraduate algebra course on the junior-senior level, and a one-semester undergraduate course in general topology.

Lectures on Algebraic Topology

by **A. Dold**
(Grundlehren der mathematischen Wissenschaften, Vol. 200)
1972. xi, 377p. 10 illus. cloth

Lectures on Algebraic Topology presents a comprehensive examination of singular homology and cohomology, with special emphasis on products and manifolds. The book also contains chapters on chain complexes and homological algebra, applications of homology to the geometry of euclidean space, and CW-spaces.

Developed from a one-year course on algebraic topology, *Lectures on Algebraic Topology* will serve admirably as a text for the same. Its appendix contains the presentation of Kan- and Čech-extensions of functors as a vital tool in algebraic topology. In addition, the book features a set of exercises designed to provide practice in the concepts advanced in the main text, as well as to point out further results and developments.

From the reviews:

"This is a thoroughly modern book on algebraic topology, well suited to serve as a text for university courses, and highly to be recommended to any serious student of modern algebraic topology."

Publicationes Mathematicae

Other Undergraduate Texts in Mathematics

Apostol: Introduction to Analytic Number Theory.
1976. xii, 338 pages.

Chung: Elementary Probability Theory with Stochastic Processes.
1975. x, 325 pages. 36 illus.

Fleming: Functions of Several Variables. Second edition.
1977. xi, 411 pages. 96 illus.

Halmos: Finite-Dimensional Vector Spaces. Second edition.
1974. viii, 200 pages.

Halmos: Naive Set Theory.
1974. vii, 104 pages.

Kemeny/Snell: Finite Markov Chains.
1976. ix, 210 pages.

LeCuyer: Introduction to College Mathematics
Using A Programming Language.
1978. Approximately 450 pages. 150 illus.

Protter/Morrey: A First Course in Real Analysis.
1977. xii, 507 pages. 135 illus.

Sigler: Algebra.
1976. xii, 419 pages. 32 illus.

Singer/Thorpe: Lecture Notes on Elementary Topology and Geometry.
1976. viii, 232 pages. 109 illus.

Smith: Linear Algebra.
1978. ix, 278 pages. 14 illus.